*...the universe...cannot be understood*
*unless one first learns to comprehend*
*the language and interpret the*
*characters in which it is written. It is*
*written in the language of mathematics,*
*and its characters are...geometric*
*figures without which it is humanly*
*impossible to understand a single word*
*of it; without these, one is wandering*
*in a dark labyrinth.*

***—Galileo Galilei***

# Mathematical Footprints

*Discovering mathematical impressions all around us*

*by theoni pappas*

Wide World Publishing/Tetra

Wide World Publishing/Tetra
P.O. Box 476
San Carlos, CA 94070

Printed in the United States of America.

1st Printing December 1999

ISBN: 1-884550-21-5

**Library of Congress Cataloging-in-Publication Data**

Pappas, Theoni
Mathematical footprints : discovering mathematical impressions all around us / by Theoni Pappas.
p. cm.
Includes index.
ISBN 1-884550-21-5
1. Mathematics--Miscellanea. I. Title.
QA99 .P377 1999
510--dc21 99-058173

*To Elvira*
*for her constant*
*encouragement and support*
*&*
*contagious enthusiasm*

*MATHEMATICAL FOOTPRINTS*

# table of contents

*MATHEMATICAL FOOTPRINTS*

With me everything turns into mathematics.
— René Descartes

# introduction

As we look around us, occasionally we see subtle impressions of the presence of mathematics. Some are current, some are left from past centuries. Tracking and discovering the trail of mathematical footprints is both fascinating and rewarding. These impressions help us understand our world and the universe, even as we discover the enormous influence of mathematics on our lives.

The first mathematical footprints date back to prehistoric times. Left on the walls of prehstoric caves are marks or symbols meant to keep track of things — perhaps the passage of time, the quantity of things gathered or bartered.

## *MATHEMATICAL FOOTPRINTS*

Over thousands of years, mathematics has revealed its influence in art, mysticism of numbers, commerce and trade, architecture, the sciences,... Mathematics is so subtle, pervasive, and necessary in our daily lives, we often are oblivious to its presence. Yet, with each passing day, mathematics is expanding its realm, placing its mark in more and more areas. Today, science could not function without mathematical tools, nor could we bank, build, travel, be entertained, enter the realm of electronics, invent, or explore the universe.

*Mathematical Footprints* follows the trail of mathematics. The order in which these mathematical footprints is presented is random, simply because that is often how they occur. For example, the Pythagorean's did not expect to uncover irrational numbers in the diagonal of a square. Nor did Fibonacci or future mathematicians expect the Fibonacci numbers to be so prevalent in nature. Or, who would have predicted fractals would become so important in describing everyday objects? History illustrates over and over that mathematical ideas have a connection to the physical world, and usually when we least expect it.

I invite you to open this book at random, and discover a mathematical footprint. Perhaps you will be left wondering, as I am, whether the first mathematical idea was born in a prehistoric cave or whether mathematical ideas are timeless universal truths waiting to be uncovered.

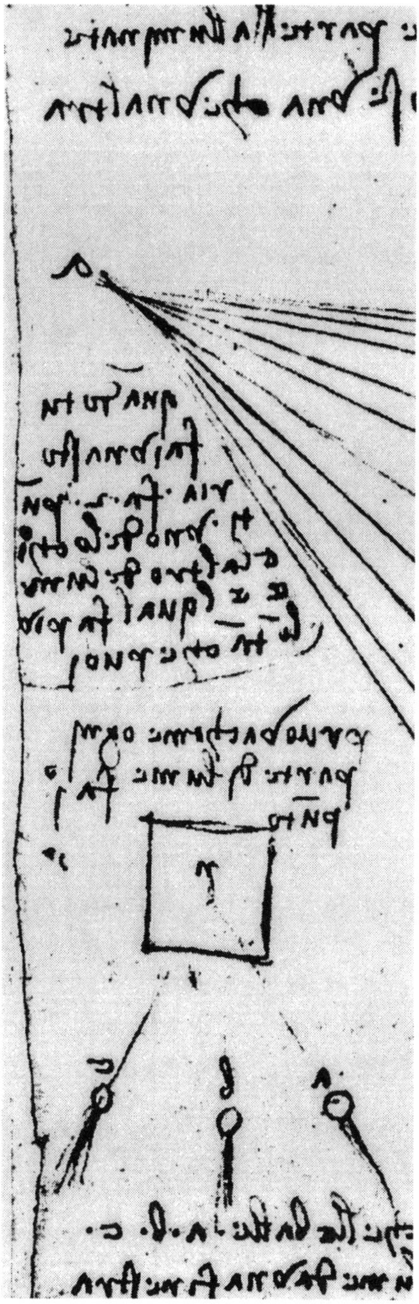

*On previous page: Work on perspective as it appears in a portion of a page from one of Leonardo da Vinci's manuscripts.*

...the ingenious method of expressing all numbers by means of ten symbols, each symbol receiving a value of position as well as an absolute value...appears so simple to us now that we ignore its true merits. But its very simplicity...puts our arithmetic in the first rank of useful inventions...

— Pierre-Simon Laplace

# early mathematics artifacts

Holding a dim light up to etched marks on the walls of caves in La Pileta, Spain, one is struck by the realization that some of these prehistoric markings are not petroglyph art or figures. Rather, they resemble something innately familiar— counting marks of some sort. What were these cave dwellers tallying? Are these examples of prehistoric mathematics? Only our imagination and intuition can supply possible answers. The earliest counting artifacts available from prehistoric times appeared in cave drawings. But that did not preclude the use of piles of stones, marks made in the dust, or marks etched in bones, in stones and on sticks. Five marks grouped together is a recurring theme found on notched bones. These early people were familiar with the five digits of their hands, so it is not surprising to find evidence of tallying by fives.

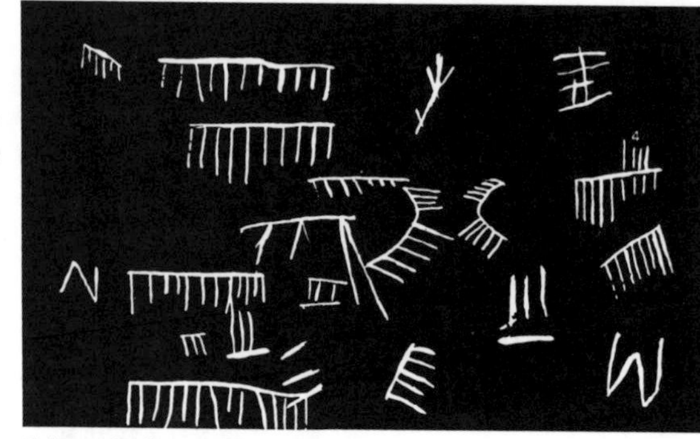
*Tally marks from the Pileta cave*

The evolution of numbers in the Sumerian civilization gives a glimpse of how abstract numbers may have evolved. The Sumerians first used clay tokens to represent commodities. Grain, for example was represented by a cone or sphere, while a cylinder shape represented an animal. As commodities shifted from agriculturally related things to merchandise, new more complicated tokens[1] in the shapes of tetrahedrons and other solids were used to keep track of such things as clothes, bread, urns of oil, worked metals. Where did they store their tokens? The agricultural tokens were sealed in "signed"[2] clay "envelopes" (clay spherical containers). These envelopes could be used as receipts or even contracts. The merchandise tokens were strung on a clay band, perhaps so they could be easily reviewed. If one lost track of what was concealed in a clay envelop, the only way to find out its contents was to break it open, thereby breaking a possible agreement. Eventually, circle indentations(for the sphere) and pointed indentations

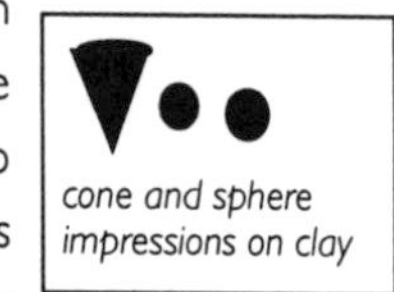
*cone and sphere impressions on clay*

---

[1] On display in the Louvre in Paris , France.

[2] Every person had a sign representing them.

(for the cone) were marked on the outside of the clay envelop to indicate its contents. Thus, the concept for a symbol gradually evolved. For example, three trees and two units of grain were first represented by three cones and two sphere tokens. These tokens were then represented by impressions on the outside of the envelop—three impressed pointed symbols for three cones and two circles for two spheres. The same recording process was eventually developed for the other tokens. As the society's function became more complicated, new recording methods evolved. From here the Sumerians went on to use a wedge shape to stand for 1 and a circle for 10. So, 15 was written ●▼▼▼▼▼. To these a larger wedge was invented to represent 60 and a larger circle for 360. Their early number system had aspects of both base ten and base sixty, yet it was not a positional system until around 2400 BC. This method of notation spread throughout the Middle East.

Ancient mathematical artifacts appear all over the world. In England we find Stonehenge(2000BC), possibly a primitive computer for predicting lunar and solar events. On the island of Crete, clay tablets were unearthed with Linear B inscriptions listing such things as farm animals, bathtubs, bronze and gold. The early Hindu numerals in the 3rd century BC had inscriptions of King As´oka (2nd century BC) which were found in a cave in Nãnã Ghãt. Numerals from the 1st and 2nd centuries were found in the Nasik caves of India. The discovery of the Rhind and Golonishev papyrus revealed mathematical work from ancient Egypt. These papyruses contained a wealth of

*How seven urns were depicted on the Pylos tablets from Crete.*

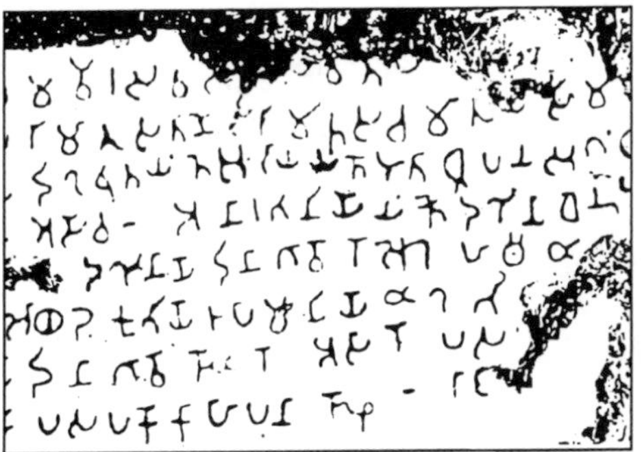

*How part of the Hindu inscription from the Nãnã Ghãt cave appears.*

*Numerals 1 2 3 4 5 6 7 8 9 10 20 30 40 50 100 200 1000*

*Brahmi numerals*
*from As´oka cave*
*circa 250 BC*

*numerals from*
*Nãnã Ghãt cave*
*circa 150 BC*

problems which used both whole numbers and fractions in their solutions. Among these are problems dealing with basic operations, linear equations, tables, practical geometric problems involving circles, rectangles, triangles and pyramids, and measurements and inventories. In a guard post along the Great Wall of China a pile of wooden sticks was uncovered, which have come to be called the Han Sticks. These sticks are from the time of the Han Dynasty (ca.200BC to 200AD) and record daily activities of the Chinese living around that portion of the Great Wall. These simple wooden sticks, among other things, reveal information about troops, their duties, payments for bricks used to build the wall, and distances of operations. The numerals appearing on these sticks show tallies written from the top down, the same direction used in China today. Instead of the traditional way, they also reveal a shortcut-like method for writing the numerals for 20, 30, and 40. These artifacts show that as the functions of a civilization become more complicated, more sophisticated mathematics is needed to handle its many tasks. Yet, artifacts need be kept in perspective, for they only furnish a peek into the past. Imagine if someone's income tax return was used as a mathematical artifact for our period. What would they surmise about our mathematics?

We see then, that as regards the fundamental investigation in mathematics, there is no final ending, and therefore on the other hand, no first beginning.

—Felix Klein

# the very pervasive golden ratio

Of the infinite number of points of a line segment, there is one very special point that seems to possess magical qualities. This point transforms the segment into a mathematical ratio[1]. This mathematical ratio is not one of your everyday ratios, but one which sets off a truly amazing chain of mathematical interconnections. It is so special that it is called the *golden ratio* (also the golden mean and golden section). First discovered in ancient Greece, the golden ratio served as a guide to aesthetic beauty and balance for the Greeks.

[1] See appendix at the back of book for an explanation on how a segment is divided into a golden ratio.

*The sides and the base of the Parthenon are golden rectangles*

Look at their architecture and you will find it; look at their sculptures and it often appears.

Once this special point is located on a segment, a *golden rectangle* can evolve. In fact, a segment's golden rectangle can generate infinitely many golden rectangles within it.

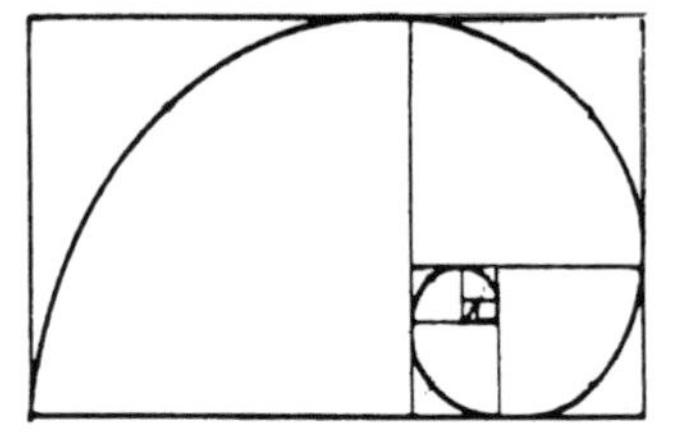

*The golden rectangle with infinitely many golden rectangles within it and its equiangular spiral*

*The nautilus shell.*

The golden rectangle with its infinitely many golden rectangles harbors a beautiful curve — the equiangular spiral — sometimes called the spiral of growth because it is found in so many objects of nature.

*The ratios AC/CD, AB/BC, AD/EF are golden ratios. The large Δs, such as ΔGEF and the small shaded ones are golden triangles.*

But the golden ratio is not confined to golden rectangles. Make a regular pentagon. Draw diagonals in it, and a pentagram is formed. The diagonals of the pentagram, and therefore the parts of the pentagram, are, in fact, divided into the golden ratio. Perhaps this is why the pentagram was thought to possess magical powers by the alchemists and the ancients.

Now look at the (five shaded) triangles formed inside the pentagram. Each triangle is called...you guessed it... a *golden triangle.* It is an isosceles triangle with base angles 72° each and vertex angle 36°. Bisecting one of its base angles, a new smaller golden triangle is formed. Continue the process of bisecting a base angle indefinitely and like the golden rectangle, the golden triangle generates within it infinitely many golden triangles. These triangles also have an equiangular spiral hidden among them.

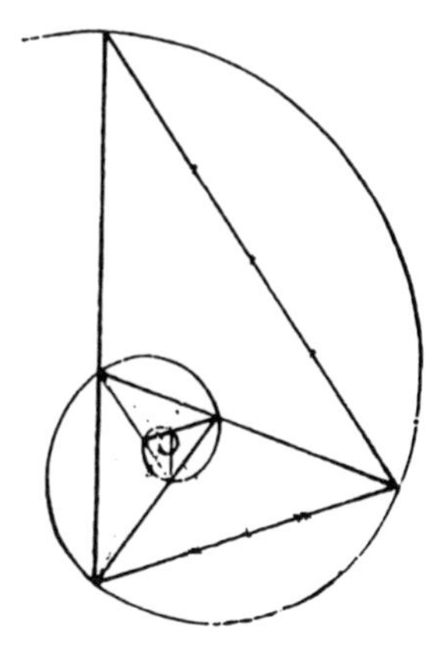

*The golden triangle with infinitely many golden triangles within it and its equiangular spiral.*

This is not the end of the chain of mathematical events generated by finding the location of that very special point on a segment. The famous Fibonacci sequence — 1,2,3,5,8,13,21...— is also connected to the golden ratio. The value of the golden ratio, designated by the Greek letter phi, $\phi$, is $(1+\sqrt{5})/2 \approx 1.6180339$ ... . Studying the numbers of the Fibonacci sequence 1,1,2,3,5,8,13,21,..., we see that each new term is the sum of the two preceding terms. The golden ratio is hidden here, in the ratios of the terms in the Fibonacci sequence. Make a new sequence that is the ratio of two consecutive terms by dividing each Fibonacci number by the term just preceding it, namely $F_n/F_{n-1}$, we get 1/1, 2/1, 3/2, 5/3, 8/5, 13/8, 21/13, ... These ratios' decimal forms are:

1, 2, **1.5,** 1.666..., **1.6** 1.625, **1.6153846,** 1.6190476, ... Notice that *these numbers alternate above and below the value of the golden ratio.* In fact, it was Johann Kepler (1571-1630) who discovered they squeeze in on the golden ratio.

But the connections don't end here. Look at the sequence $\phi$, $\phi^2$, $\phi^3$, $\phi^4$, $\phi^5$, ... When it is expanded using $\phi$'s value $(1+\sqrt{5})/2$, it becomes $\phi$, $2\phi+1$, $3\phi+2$, $5\phi+3$, $8\phi+5$, $13\phi+8$, ... and again the Fibonaaci terms are linked to $\phi$.

$$\phi = \frac{1+\sqrt{5}}{2} = 1.615\ldots$$

What other geometric objects have the golden ratio hidden in them?

- if a *regular decagon* is inscribed in a circle, the ratio of the radius of a circle to the side of the decagon is the golden ratio
- if the edges of a *regular octahedron* are divided into golden ratios, these 12 points are the vertices of a *regular icosahedron*.
- the twelve centers of the 12 faces of a *regular dodecagon* can be joined to form three mutually perpendicular golden rectangles.
- the vertices of a *regular icosahedron* can be joined to form three mutually perpendicular golden rectangles
- the golden ratio appears in the formation of *Penrose tiles*

It is astounding to discover the various mathematical connections that arise from locating a point on a segment. Look further and the golden ratio's footprint is bound to appear elsewhere.

# seeing is not believing

*mathematics & optical illusions*

No great artist ever sees things as they really are. If he did he would cease to be an artist.

—Oscar Wilde

There is a definite distinction between what is before our eyes, and what we think we see. How the mind's eye changes reality was known by the ancients. In fact, the Greeks specifically altered the columns in their buildings to make them appear straight to the viewer. The architects of this time knew that structures built perfectly straight did not appear straight. This is because the curvature of the retina affects how we see things. As a result, the columns of the Parthenon and its rectangular base were made slightly convex.

Where precision is necessary we cannot rely solely on our eyes, but must resort to measuring — one of the earliest functions of mathematics. But this is not the only function or influence mathematics has on what we see. Renaissance artists of the 15th century discovered how perspective and concepts of projective geometry could be used to enhance the dimensionality and realism of their works. The Renaissance artists were not the first to try to portray a 3-dimensional image. In a cave at Lascaux some 15,000 year ago, a cave dweller — trying an innovative technique—attempted to capture the mood of 3-dimensions by specifically drawing on protruding stones rather than on a flat wall.

The study of the power of optical illusions really flourished in the 1800s when astrophysicist Johann Zollner (1834-1882) stumbled upon a piece of fabric with a design that made parallel lines seem nonparallel. Naturally, such discoveries were very important in astronomy, which relied so heavily on observations. As a result, the scientists[1] of the 19th century studied, explained and categorized the various illusions created by manipulating marks, space, shapes, perspective, and negative images to determine how our minds were tricked into seeing something that did not really exist. It was also at this time that *stereoscopy* was developed which created 3-D images from 2-D images. Sir Charles Wheatstone invented the stereo viewer in 1838, and the following year photography was invented. The two played off one another. The stereoscope made its public debut in the London at the Crystal Palace Exhibition in the mid 1880s. A *stereograph* relies on the images captured by our eyes from their slightly differing perspectives. Even artist Salvador Dali was so enthralled by the 3-D images created

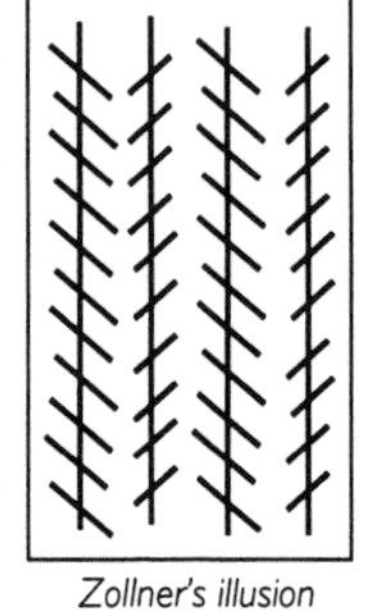

*Zollner's illusion*

---

[1]Among these were physiologist, physicist and mathematician Herman von Helmhotz (1821-1894), physicist and psychologist Ewald Hering (1834-1918), physician Johannes Muller (1801-1858), geologist, paleontologist and stratigrapher Albert Oppel( 1831-1865), philosophy professor and scientific psychologist Whilhelm Wundt (1832-1920).

by stereocopes that he painted several works in the 1960s and 1970s that were to be viewed with the stereo viewer. Today, optical illusions of the past can be thought of as non-computerized virtual realities.

*Stare at these two figures, and let your eye muscles relax. One will move into the other, and then the two squares become 3. Do all three exist?*

Using the science of the eye coupled with mathematics of perspective, dimensions, and geometry, optical illusions are continually being developed. A major breakthrough took place in 1979 when Christopher Tyler, psychophysicist, and programmer Maureen Clarke presented the first single-picture stereogram. Later, random dot stereograms were created. These were formed from two identical dot pattern images (one for the right and left eye) that were placed slightly out of sync and then covered with a spread of random dots. In the 1990s, the concept of the single-picture stereogram was popularized with the "magic eye" posters and books. Stores at shopping malls across the country had window displays of these stereograms. Groups of people would gather in front of a window display and stare at the posters, trying to capture the hidden 3-dimensional image. Without using any apparatus other than one's eyes, it is fascinating to experience 3-D images suddenly appearing before your eyes off a flat page. Viewing one of these single-picture stereograms (also referred to as 'magic eye' pictures) is similar to discovering an invisible dimension— a hidden world right in front of your eyes. Today, the computerized worlds of virtual reality using the techniques of stereograms create their images by having a computer display for each eye. All made possible by optical illusions joining forces with the mathematics of multiple dimensions and 3-D graphics[2].

---

[2] In 1843 Sir William Hamilton introduced 4-dimensional numbers, which today have proven invaluable for creating 3-dimensional computer graphics.

Yet, understanding and using optical illusions is more than a pastime or a curiosity. Optical illusions are being used in conjunction with highway safety. Using the studies of optical illusions, traffic engineers are experimenting with the use of bold patterns painted on dangerous stretches of roads "to trick" motorists into modifying their driving. These bold chevrons, bars and stripes appear equally spaced, but in fact are shrinking, and give the illusion that one's speed is increasing. The driver reacts by slowing down, and trying to maintain the reoccurring stripes at equal intervals. Calculating the frequency of the interval of stripes and the types to use is crucial to reaching the desired effects. Whether such optical tricks will have lasting effects on drivers who frequent the optically treated road has yet to be determined.

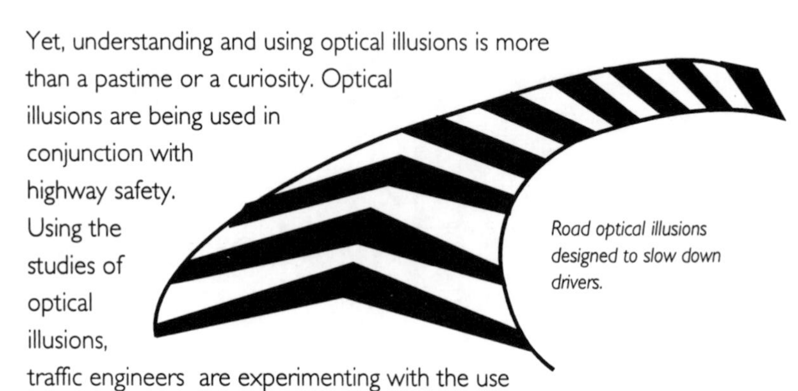

*Road optical illusions designed to slow down drivers.*

And so the study and development of new optical illusions and their uses continue, and intrigue the layperson, the psychologist, the artist and the scientist. In the near future, perhaps these studies will reveal important knowledge about the function of the eyes and the brain. As Salvador Dali said, "To gaze is to think."

Cyberspace doesn't have all the physical constraints. There are no walls...[a] scary, weird place, and with identification it's a panopticon nightmare.

—David Chaum

# mathematics & your money

## *electronic money & cryptography*

*Two bits* is commonly used slang for a quarter, and *four bits* for a fifty cent piece. Little do those using these terms realize that eventually all denominations of money will be bits — that is digital bits— 1's and 0's. We have entered a new money era. Just as shells and beans were replaced by coins and coins by bills and bills by checks and checks by credit cards and ATM cards, now we have been introduced to electronic money, e-money and virtual banks. Some are skeptical and fearful. Can we trust this new currency and banking? Will keeping track of it be a headache? What about counterfeiting? Will we be able to retain the privacy of using cold cash? We are on the threshold of this change. It won't be fully realized for some years, but it is happening gradually. Slowly we have become accustomed to electronic transactions and electronic coding. Nowadays there seems to be a number for everything. Almost every item we purchase in stores has a bar-code. The government identifies us via our social security numbers. Each car has a license number. You may have a PIN (personal identification number) to access calling cards or ATM cards. Even our physical attributes are being mathematized using the science of biometrics by such things as eye scans and the geometry of hands. There are — access numbers for going on-line—credit card numbers—health card numbers. Gradually we are becoming conditioned to multi-faceted types of e-data and e-money.

**How did virtual banks get started?**

Banks already transfer huge sums of money electronically on a daily basis. They would like to be able do all transactions electronically. Can you imagine how much banks would save by not having to physically handle and move armored carloads of money? Electronic money has already begun to redefine the banking industries' job force, and more changes are on the way when digital cash comes into widespread use. One thing is certain, mathematics will play an important role in our pocketbooks. What role? Securing your money. Who would have believed that cryptanalysis, cryptology, and cryptography would be in such demand? With encryption businesses growing and techniques and hardware being constantly developed, revised and updated, mathematical research in these fields is constantly expanding. More and more people are being lured to on line virtual banking, either by their traditional banks offering Internet banking or by actual banks, such as NetBank, TeleBank, CompuBank and WingspanBank.com. Most people who do Internet banking also have traditional accounts because as of now no easy methods have been devised by a virtual bank to make deposits, no virtual ATMs exist for cash on demand, and there aren't any virtual safe deposit boxes. What are the

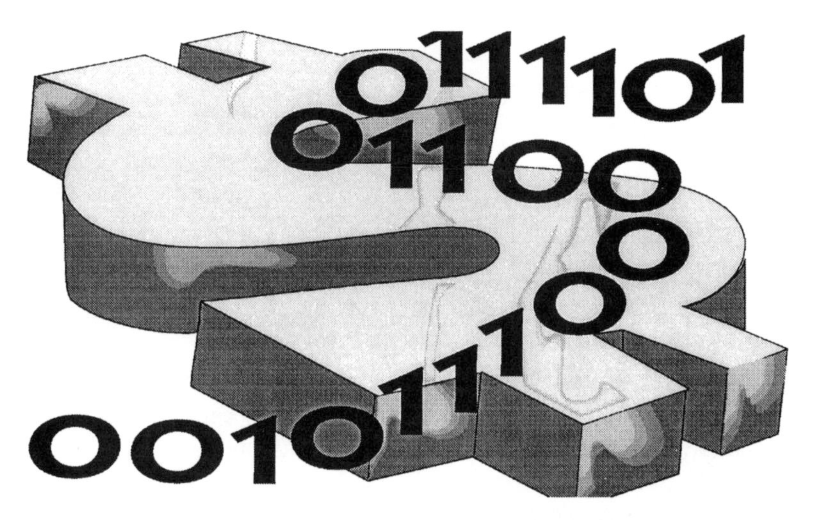


pros and cons for virtual banking? *Pros:* Convenience (you do your banking from home), speed(you don't have to wait in a bank line), 24 hour a day access, higher interest rates on savings and lower fees. Since each Internet transaction costs a few cents while an actual bank's cost is around $1.50, virtual banks can afford to pay more in interest. *Cons:* No personal interface or physical place to get information or complain to an actual person.

**What are some of the safeguards developed thus far for securing digital information and thereby e-money?**

***"key-escrow"** encryption technology* — devised to insure privacy by using an integrated circuit chip which encrypts data and speech using a classified mathematical formula. This technology would also allow access to information by authorized law enforcement agents possessing a master "key". Some flaws in this method are currently being explored.

***public-key cryptography*** —a method whereby the sender and receiver of electronic data use different mathematically related "keys" to verify one another's authenticity, thereby allowing strangers to communicate or carry on business transactions. The sender of electronic data uses one secret mathematical "key" to encrypt a message and the receiver has a different mathematically related key to decipher it. Each has both a private and public "key".

***private-key cryptography*** — with this system both the sender and the receiver hold identical mathematical "keys".

***digital signatures*** — a method for establishing the authenticity of the sender and the integrity of the electronic data, but not used to encrypt the information.

***blind signatures*** — a method to authenticate a number or data and its integrity while being unable to trace or identify the sender.

***What will be the form of this electronic cash?***

Even as banks and businesses transfer money electronically between banks and accounts, so do individuals transfer electronic money between their linked accounts and instruct banks to pay certain bills. There are no delays in being paid or paying bills because electronic "checks" clear at the speed of light. The ideal scenario is that you would keep track of your assets electronically. In the same way banktellers pull up your accounts on a bank computer monitor, you are able to do the same from your home computer. You can instantaneously check if what you requested was actually paid, transferred, received and recorded. You can print paper statements with your printer if you so desired. In the future, if you want cash, you will be able to download x-dollars onto your cash cards using your computer or electronic wallet, remembering that like cash it does not earn interest. Like cash it will have the versatility and speed of use for a variety of purchases, be they gasoline, restaurants, bridge tolls, or groceries.

***What form will cash cards take?*** Hopefully, the options of traceable and untraceable cash cards will be available.

•***traceable cash cards*** — A traceable cash card attached to a PIN is one option. By using this card, all your transactions, everywhere and when and how you spend your money are a matter of record—including where and when you parked your car, paid your overdue library book fine, where you stopped for coffee. In essence, a daily trail is left by your expenditures and is open to public scrutiny. Granted, you won't need to keep track of your expenditures. The IRS would be able to access your information and prepare your tax return.

•***untraceable cash cards*** — If on the other hand you want to retain your privacy, you could opt for an anonymous cash card, and as with cold hard cash, you retain the privacy of how, when and where you choose to spend your money. No one could trace your daily electronic steps, short of hiring a detective to follow you.

In both cases, you would no longer need to fiddle for parking meter change or carry a cumbersome wallet with bills and coins. Instead you would have an electronically encoded card with a dollar amount expressed in digital bits, a mathematical encryption "key", and an electronic signature, which can be a digital signature or a blind signature.

***Where is electronic cash in use today?***

Digital cash cards have been evolving in many forms and with many names—the *debit card*, the *smart card*, the *electronic purse*, the *virtual wallet* or *cash*. One predecessor has been the electronically encoded transit card purchased for a specific amount to use on subways and buses. In Denmark, Portugal, Singapore and Britain, cash cards by various businesses are currently being introduced. Many businesses[1] are beginning to tap and explore the potential of this new area. As yet the Federal Government has not expressed interest in exploring, monitoring or regulating electronic cash. It is only a matter of time, however. The logistics and success of such a monetary system rests on the effectiveness of the mathematics and cryptography behind it.

---

[1] Among these businesses we find DigiCash, CyberCash, VISA, Citibank, Mondex, Microsoft, RSA Data Security.

Architecture is akin to music in that both should be based on symmetry of mathematics.

—Frank Lloyd Wright

# mathematics–
## *the framework of architecture*

Mathematics has influenced the shape and integrity of every building we enter, every bridge we travel over, every tower we see. These mathematical architectural footprints date back thousands of years. The Egyptians used mathematics to design their massive pyramids. They knew about the concepts of the Pythagorean theorem and how they worked. They used Pythagorean triplets such as 3, 4, 5 along with ropes stretched to these lengths to form a right triangle and furnish the needed right angle for the base of their pyramids. The Babylonians solved many problems involving square root answers. And the Greeks proved the Pythagorean theorem and the

*Oracle buildings use familiar shapes enchanced with high tech facades, Redwood Shores, CA.*

existence of the irrational numbers it generated. Incorporating these ideas along with the ideas of the golden rectangle and concepts of optical illusions, the Greeks constructed their famous Parthenon. But mathematics' role in architecture did not stop with ancient times and the Roman arch, nor did it stop later on with works of Christopher Wren, or Leonardo da Vinci, or the architects of Gothic Cathedrals, the Byzantine mosques, the Taj Mahal, the Mayan pyramids, or the Forbidden City. Today, these past mathematical footprints, along with new ones, are defining contemporary and futuristic architecture. The influence of the past, coupled with new mathematical ideas, technologies and materials, give architects the tools to explore and create new shapes and awe inspiring structures.

All that we see or seem is but a dream within a dream.

—Edgar Allen Poe

# hyperspace & beyond

In 1854 Georg F. B. Riemann was asked by Karl Gauss, his professor, to give a lecture to the faculty at Göttingen on the foundations of geometry. The concepts he formulated would have a profound impact on the evolution of mathematics and physics. He presented ideas of a new geometry, a non-Euclidean geometry, in which Euclid's 5th postulate[1] was replaced with Riemann's postulate[2]. This geometry had its own special properties[3]. Riemann also spoke of ideas of n-dimensional space. He built on Gauss' analogy of a world which exists on a flat sheet of paper on which 2-dimensional bookworms reside. He proposed warping this universe by crumpling the sheet of paper. The worms would still crawl over this crumpled paper, unaware that their world was no longer flat. At times they would experience forces (i.e. wrinkles in the paper) which would steer them from a direct path. Continuing along this line, he proposed an analogous situation for our 3-dimensional world which was described as crumpled (warped), but in the 4th dimension. Here the forces such as electricity, magnetism and gravity are the wrinkles of the crumpled hyperspace[4] that guide us in various directions. In his lecture, he even dealt with multiple connected spaces, i.e. wormholes[5]. In the 19th century, Riemann's talk flung open a door to a strange new world. Just knowing that one might be engulfed by an invisible world, a world with so many more facets than ours, it is no wonder the quest for a glimpse of it has attracted a diverse group of people. Among these we find Herman Hemholtz, Pablo Picasso, Johann Zollner, Lenin, Charles Hinton, Salvador Dali, H.G. Wells, Oscar Wilde, William Crookes, J.J. Thompson, Dostoyesvsky, Edward Abbott, Diego Rivera, and many others.

---

[1] Euclid's 5th postulate, the Parallel postulate states: *Through any point not on a given line there is only one line parallel to the given line.*

[2] Riemann's parallel postulate states: *Through any point not on a given line there are no lines which can be drawn parallel to a given line.*

[3] Other changes by Riemann include the interpretation of another postulate of Euclid, *a straight line may be produced to any length in a straight line.* His postulate read *a line is boundless but not infinite.*, i.e. it has no ends but is finite in length. Such a geometry is consistent with a geometry that exists on a sphere, where all lines are great circles. On a sphere any two great circles always intersect in two points and thus no lines are parallel. In addition, on a sphere the sum of the angles of a triangle total more than 180° and as a triangle's area increases the sum of its angles increase.

[4] Mathematically speaking, hyperspace is the fourth dimension—the spatial dimension after length, width and depth. In physics and as described by Einstein, time is often referred to as the fourth dimension. Thus, physicists refer to the mathematical fourth dimension as the fifth dimension, 4-spatial dimensions plus time.

[5] To understand wormholes, return to the booksworms' world and this time consider a stack of papers which are slit at different places and joined at the slits. A worm could pass from one sheet of paper (its universe) to another by slipping through the holes (the slits) connecting the sheets.

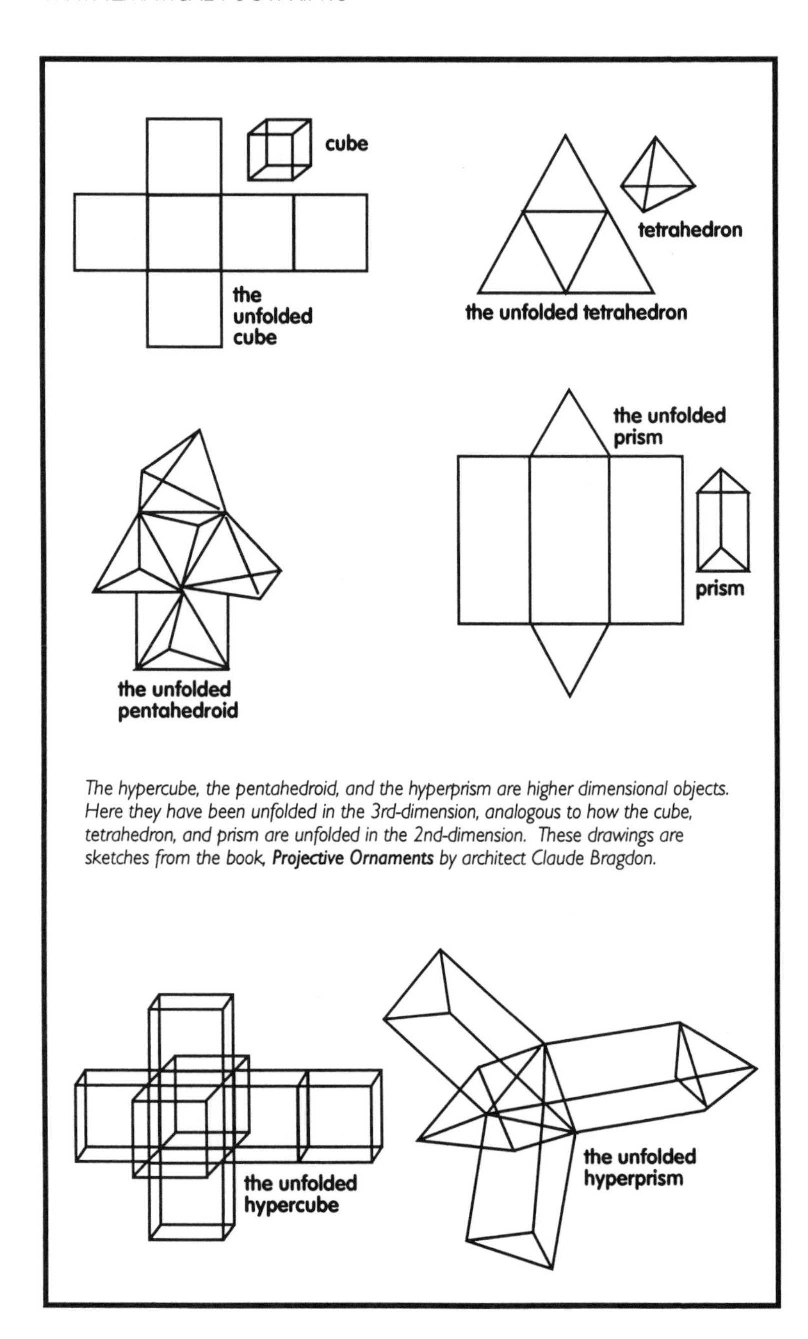


*The hypercube, the pentahedroid, and the hyperprism are higher dimensional objects. Here they have been unfolded in the 3rd-dimension, analogous to how the cube, tetrahedron, and prism are unfolded in the 2nd-dimension. These drawings are sketches from the book,* ***Projective Ornaments*** *by architect Claude Bragdon.*

Mathematicians, artists, scientists, spiritualists, mystics, philosophers discussed, debated and tried to visualize higher dimensions. Since mathematics does not require concrete examples of such worlds, it was possible to develop the mathematics of elliptic geometry along with higher dimensions. But, our 3-dimensional world made it impossible to carry out experiments of this invisible world. Consequently, at this time physicists, on the whole, abandoned his ideas. Fearful of ridicule, they did not openly consider higher dimensional worlds, because that which could not be measured was not considered good scientific inquiry. Riemann's ideas evolved into things currently found in science fiction.

Today, there is a renewed interest in and a renaissance in the study of higher dimensional space. The mathematics of higher dimensional space is playing an essential role in explaining a new view of cosmology. Higher dimensions are providing theoretical physicists with tools and new perspectives to explain perplexing questions and conflicting theories[6]. For example, without the fourth dimension there is no explanation of how a light wave can travel through a vacuum. Light is considered both a particle and a wave. Waves need a medium in which to travel. Since a vacuum is a void how can a light wave pass through it? By considering light as a vibration of the fourth dimension. Theoretical physicists are actually using higher dimensions to explain ideas that had eluded them for decades. Unifying field theories, Big Bang, TOE (Theory of Everything), multiple universes, time travel, wormholes, parallel universes, cluster universes, the 10 and 26 dimensions of Superstring theories[7]— are all

---

[6] For example, theories of Einstein and quantum physics.

[7] A version of the Superstring Theory contends that the building blocks of all forms of matter and energy of the universe are infinitesimal strings. Prior to the Big Bang the universe was 10(or 26) dimensional — nine spatial dimensions and one time dimension — but most unstable. Hence, the Big Bang occurred and the universe expanded in only three spatial dimensions. It is continually expanding. The other six dimensions remained entwined and encased in compact geometries that measure $10^{-33}$ cm, across. The minuteness of these things explains the existence of the many subatomic particles that are continually being discovered by atom smashers. Furthermore, these strings vibrate and produce frequency, much as a string instrument does. The notes produced by the vibrations are believed to define all forms of matter and energy.

evolving from mathematics that originated over a century ago. These are not isolated studies, but widespread ones, emanating from research labs around the world, where countless papers have been written on theories utilizing higher dimensions.

We may be entering an era in which many profound ideas and theories will not be measurable other than by the intellect. That is where the role of mathematics comes into play. Although one thinks of mathematics as providing tools for measurement, mathematics also provides tools to describe and measure the immeasurable. Although we cannot visualize higher dimensions, mathematics has created these dimensions on paper. Through mathematical equations these ideas exist. They appear in formulas, expressions, and geometries. And they provide the means and the tools for the scientific exploration

I paint things as I think them, not as I see them.

—Pablo Picasso

# mathematics & cubism

Around 1907 Pablo Picasso and Georg Braque introduced a new style of art which was adopted and enhanced by other artists through the year 1914. Among these artists are Albert Gleizes, Jean Metzinger, Marcel DuChamp, Francis Picabia, Fernand Leger, and Juan Gris. This art came to be called *cubism* when art critic Louis Vauxcelles commented, "*These new works look like a bunch of little cubes.*" But his description was misleading. These new works were not just a conglomeration of cubes, but an entirely new way of depicting an object. In the words of Picasso *"I paint things as I think of them, not as I see them."*

*Man with Violin by Pablo Picasso. (c. 1910-1912)*
*Philadelphia Museum of Art: Louise and Walter Arensberg Collection.*

What ideas were occupying the artists' thoughts when painting in this new avant garde style? They were not using cubes just to represent objects in their paintings or to carry out Cézanne's belief that *"everything in nature adheres to the cube, the cone and the cylinder"*. The cubists

*Nude Descending a Staircase by Marcel Duchamp. (1912) Philedelphia Museum of Art: Louise and Walter Arensberg Collection.*

were going beyond using these basic geometric objects as models; they were delving into a new reality that described objects in an entirely new framework — the framework of a new dimension and a new sense of time. At the turn of the 20th century, the introduction of the ideas of non-Euclidean geometries, higher dimensional mathematics, and the space-time dimensions of Einstein were making their impression on the world.

The higher dimensional space presented by Riemann in his elliptic geometry intrigued a cross-section of society. Discussions of higher dimensional worlds were not restricted to mathematicians, but even sparked the interest of politicians, artists, poets, scientists, laypeople. In the early 20th century the mathematical ideas of the 4th-dimension and higher, of Einstein's 3-dimensions and time, and of strange non-Euclidean worlds came to the attention of a group of artists in France. These ideas were discussed at Saturday evening gatherings at Gertrude Stein's salon on Reu de Fleurus.

Here mathematician Maurice Princet held forth on the ideas of this new geometry. It was also at this time that the well known French mathematician Henri Poincarè conceptualized and described certain non-Euclidean worlds in his various works. These discussions and writings fueled public interest and in particular the artists' curiosity, and contributed to the emergence of cubism. Perhaps these were some of the ideas that were occupying the thoughts of Picasso and Braque as they painted.

It was not the first time that geometry would influence art. Geometry, be it Euclidean or non-Euclidean, is considered a conceptual form of mathematics. It deals with physical objects, objects we visualize and use to fill space. Artists transform or fill space in the process of developing their works. Representing space requires defining its dimension. During the Byzantine era, the religious art of icons was created in a 2-dimensional manner. The introduction of projective geometry during the Renaissance helped the artists capture the realism of the 3-dimensional world. The elements of projective geometry — point of projection, parallel converging lines, vanishing point — allowed the artist to transform the 2-dimensional canvas into a 3-dimensional world. But even as elliptic geometry delved into higher dimensions, so too were artistic works not to be confined to the 3-dimensional world. And so this new geometry ushered in a new style of art.

In cubism the use of different geometric forms placed at various angles convey the multiple facets of an object and thereby transform it into a hyperobject[1]. Even as the rendition of the hypercube simultaneously shows all eight cubes composing it, the cubist can simultaneously show all aspects ( front, back, sides, tops, bottoms) of the objects without adhering to a "logical" sequence. In Picasso's *Man and Violin,* our eyes may try to follow the edges outlining an object, but our mind's eye has trouble because the path is unlike any we have experienced in our 3-D world.

---

[1] In mathematics the prefix *hyper* connects the word to which it is attached to the 4th-dimension, e.g. hypercube is the 4th dimensional counterpart of the 3rd-dimensional object, the cube.

An object in a cubist's painting may possess many profiles or views, each of which could depict a different time frame. This is the case with the violin and its parts that appear many times in different places of the canvas, as do multiple profiles of the man. Similar experiences are in store for the viewer in Braque's *Man and Guitar.* In Léger's *Three Figures,* one is captivated by the multifacets of the figures and the movement of the bodies. The eye does not know where to begin to view the work. While in Marcel Duchamp's *Nude Descending a Staircase,* all sequences of the figure moving down the stairs occur simultaneously.

In cubism, space and time seem to work as one unit rather than time being shown as a progression of events in space. Unlike the realism of Renaissance art, cubism makes the mind sort out and work with the objects presented, since they are unlike any in our world. Thus, cubists' paintings do not necessarily follow the prescribed sequence of time as we know it, but instead bring to view visual representations of new worlds, multiple perspectives, embodied in multiple periods of time. A cubist can depict different dimensions, different moments in time plus higher dimensional worlds. Although the cubist period per se was short lived, its impact on modern art has been dramatic.

The mathematical phenomenon always develops out of simple arithmetic, so useful in everyday life, out of numbers, those weapons of the gods: the gods are there, behind the wall, at play with numbers.

—Le Corbusier

# le grande arche

*a building with a brain*

A panoramic elevator carries you through "the clouds" to the top of the Grande Arche at La Defense Center in Paris, France. The simple yet elegant cube is the model used for this architectural wonder. Although the Grande Arche is based on this classic shape, its unique and elegant design has a futuristic quality to it. Yet this is no ordinary cube, but a hollow or open cube, resembling the form of a tesseract with sides about 360 feet long.

The Grande Arche posed greater obstacles than a normal cubic building; but its aesthetic rewards

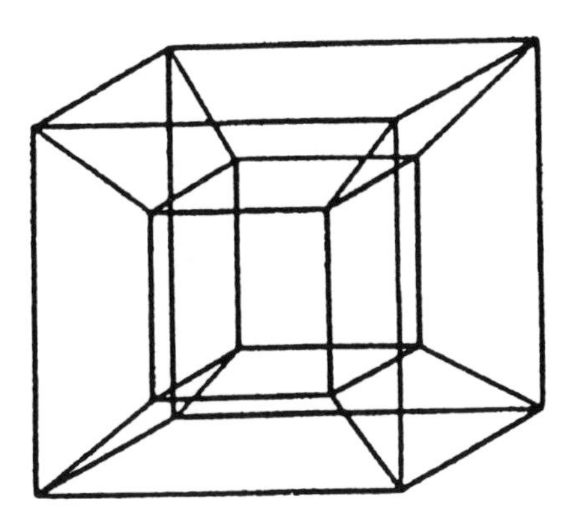

far outweigh the problems that had to be solved in its construction. It could not be constructed story by story because of the void that had to be left in its interior. Consequently, it is embedded with a dense iron skeleton which is reinforced every 21 meters by huge vertical support beams and a web of cables and pipes running throughout the structure. It had to be engineered with prestressed high strength concrete[1] formulated with high resistance and pumped up to the various levels.

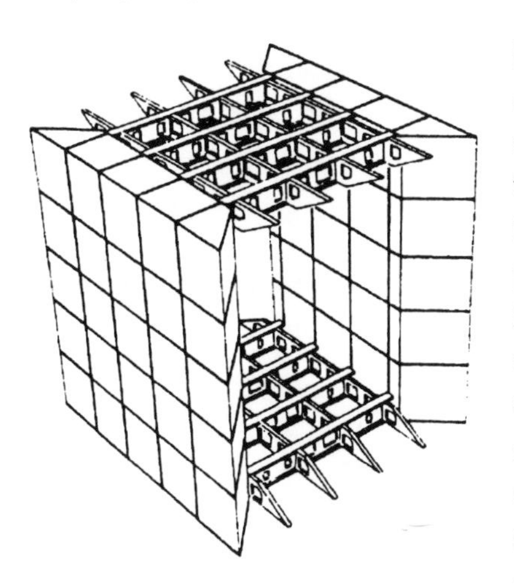

Each of its subframes is a "mini" cube and an independent seven story hollow structure transmitting its load to the frame. The cube rests on 12 pillars, each of which must bear as much as 30,000 metric tons. Because of the massive weight of the structure, extensive initial calculations and estimates had to be considered to account for settlement of the structure on its limestone base. For each phase, vast amounts of data were entered and analyzed using

---

[1] For the average bridge, 264 pounds of concrete per cubic yard of steel is used. The Grande Arche has as much as 750 pounds per cubic yard of steel.

*The Grande Arche at La Defense, Paris, France*

computers. Multi-dimensional mathematical modeling programs were devised to solve the various engineering problems. Unlike the pyramid, the cube is a more unstable shape, so the problem of containing the force of pressure with its bevelled angles had to be solved. The top or roof (also referred to as the bridge) of the structure is designed to support up to 1 metric ton per square meter. The base of the cube rests on limestone 35 meters below the surface.

How can any damage or structural problems be detected as the Grande Arche ages? This incredible structure is wired to three independent computers which monitor its movements or problems. For example, hooked to each of its 12 pillars are 150 electronic auscultators, which are able to detect the slightest movement or vibration. If a pylon becomes ensconced by even a few millimeters, the pylon's neoprene cushion (four inches thick) can be be replaced by a 2,000 ton hydraulic jack which would rectify the situation. Because of sophisticated high-tech computerized sensing systems throughout the entire structure, the Grande Arche is often referred to as "an intelligent building". This intelligent building has its information monitored and relayed by independent computer controlled robots strategically situated throughout the structure. Don't let its name be misleading. It is not only an arch, but is a building with offices, conference rooms, a theater, gift shop, etc.

Whose brainchild is this structure? In 1982 an international architectural competition was launched by the French government to find a design that would be both the focus and culmination for La Defense center. The Grande Arche was designed by Danish architect Johan Otto von Spreckelsen[2]. He situated it along the historic axis (thoroughfare) of Paris, which aligns the Louvre, the Place de la Concorde and its Obelisk, the Arche de Triomphe, and the Champs-Élysées, thereby making the great monuments of Paris part of the vista from the roof of the Grande Arche. In addition, Sprekelsen purposely located the Grande Arche 6° off the axis in order to give the structure a 3-dimensional perspective quality when viewed along the thoroughfare from the Arche de Triomphe. "Clouds" of synthetic materials[3] are constructed in the hollow area of the cube to give depth to the space's interior. The Grande Arche is an aesthetic, intellectual and architectural wonder.

---

2 Spreckelsen appointed French architect Paul Andreu to execute the plans for his design.

3 Cables and bars were used to frame the fiberglass material that was prestretched and coated with teflon.

Men have become fools
of their tools.

—Henry Thoreau

# the mathematical pandora's box

The quest by humans for new and improved tools has been going on over the centuries. As problems involving numbers, operations, and counting became more and more complicated, tools to do tedious computations more and more quickly with improved accuracy also evolved. We have progressed from notched bones and sticks, knotted ropes and the abacus to today's personal computer. The journey has been varied and fascinating. Around 1614 John Napier introduced logarithms and his famous Napier bones (a box of calculating rods), which seaman and merchants carried to help with involved computations. In the years

*A rendition of Joseph-Marie Jacquard's loom and the portion of one of the punched cards used to program the loom.*

1620 and 1621, British mathematicians Edmund Gunter and William Oughtred, using logarithmic concepts, were instrumental in devising prototypes for the first slide rules. The mechanical calculators of Blaise Pascal (his 1642 adding machine) and of Gottfried Leibniz (his 1673 calculator which added, subtracted, multiplied and divided) were invented to help human computers with the monotonous work of computation. Another major breakthrough surprisingly occurred in the weaving industry. In 1801, Joseph-Marie Jacquard invented a special loom which he designed to receive its instructions via a program provided by punched cards Twenty years later, Charles Babbage used such cards in the design of his Analytical Engine to instruct the mathematical operations his machine would perform. Unfortunately his difference and analytical engines were never realized during his lifetime, but his plans sparked new ideas about what tasks such machines could be made to do

The 20th century is responsible for opening Pandora's box. In 1907 Lee

De Forest invented the vacuum tube, which eventually replaced mechanical levers and gears in calculators. The first operating computer that functioned with vacuum tubes was invented in 1942 by John V. Atansoff and Clifford Berry. The ABC computer (Attansoff Berry Computer) was designed to store its binary data in its capacitors, which led to a new and important use for binary numbers (the 0s and 1s represented the OFF-ON states of electricity). From now on, the term *computer* would no longer refer to a person who did computations for a living, but, rather, to a machine which performs calculations and many other functions. The modern day (silicon chip, integrated circuit, solid state) computer literally revolutionized how we do everything — from medical exams to making telephone calls to posting things we buy with plastic instantaneously to tracking criminals to describing the spread of diseases to handling money to drawing graphics to creating movies to scientific experiments to the impact on advertising and communicating via the Internet. The computer has become both the answer and the cause of many problems. For most individuals its power exceeds what they can use or handle. Does this stop the never ending quest for more computer power, more computer memory, more computer uses? No. As in the Greek myth, when Pandora's box was opened, it was impossible to return its contents. What does the 21st century hold in the realm of computers? Look for mathematics to unleash the possibilities of the quantum computer, the molecular computer, and the optical computer. Look for fuzzy logic to be tied to artificial intelligence and soft computing which will revolutionize how and what tasks a computer can perform. Look for computer modeling and cellular automata to be used in exploring diverse universes. Look for the invisible computers of nanoworld to attack tasks on the atomic level. Look for the computer to morph itself into almost any tool. — And if anything goes wrong, look for the computer scapegoat.

# mathematics & the body

Mathematical analysis is as extensive as nature itself.

—Joseph Fourier

Today, we are realizing that mathematics has left significant imprints on the sciences of anatomy and medicine. For centuries, scientists and physicians have sought to develop methods to cure illnesses and understand the functions of the body. Identifying the divine proportion in the human form and analyzing spirals in the various formations of the body are some early notions used to mathematize the body. The ancient Greek sculptor Phidias used the divine proportion in his sculptures. Later, such artists as Leonardo da Vinci and Albrecht Dürer searched to quantify the shape of the human body. Galileo made note of the clockwork of the human heart rate by comparing it to the rhythm of the pendulum. Today, the mathematics of both the form and the functions of the body is ever growing. It has gone far beyond seeking numerical norms and ratios for bodily functions such as — measuring blood pressure, cholesterol levels, ratios of LDLs, HDLs, using EKGs and MRIs. The medical world is now relying on a vast spectrum of mathematical concepts, calculations and inventions to try to unravel mysteries of the human body. Among these we find—

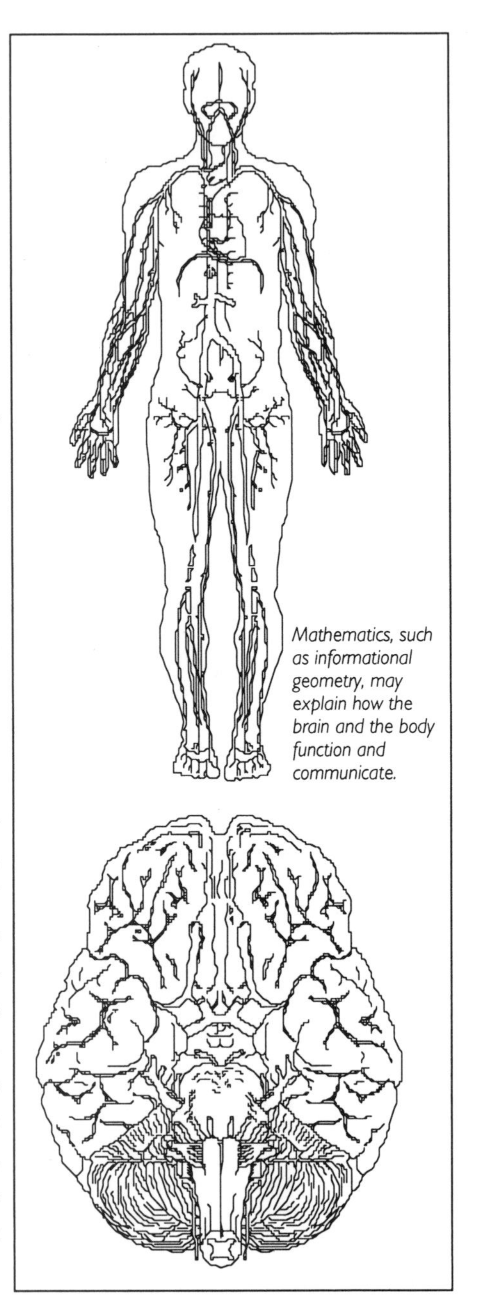

*Mathematics, such as informational geometry, may explain how the brain and the body function and communicate.*

- *Spiral scan machines* are used to produce continuous helical CT scanned images. In the past, machines produced only a single cross-section CT scan, but this new machine is designed to scan along a 3-D spiral, thereby giving more complete and uniform information.

•*A map of the hereditary information found in the 23 pairs of human chromosomes* is being compiled by teams of geneticists working with computers to create an atlas of the base pair sequences for all 80,000 to 100,000 human genes. The Human Genome Project is due for completion around 2003.

•*Load pressure on joint motion* of soft tissue is being calculated using geometry and computer simulations.

•*Chaos theory and complexity* are being explored in connection with heart rhythms, brain activity, and seizures.

•*Computer techniques, imaging and modeling* are being used to fill in gaps, just as the human brain fills in its blind spot by surmising surrounding information.

• *Fractal geometry* can now be used for early detection of osteoporosis. The composition of bones resembles a honeycomb structure. Fractals are used to describe growing or changing matter, since they themselves are ever changing. If a bone's honeycomb structure begins to deteriorate, this degeneration can be detected by determining if the bone's fractal dimension has decreased by using a digitized X-ray or magnetic resonance image and computer technology. These fractal dimensions will indicate the onset of osteoporosis long before bone mass decline can be detected. This improved method for measuring the bones fractal dimensions was devised by Raj Acharya of State University of New York at Buffalo. To arrive at the bone's fractal dimensions, Acharya formulated nonlinear algorithms based on mathematical morphology. In addition, fractals are currently used to describe ion channels and movements through these channels.

•*Binary optics* is being utilized to produce specialized optic pieces.

•*New mathematical shapes* are being created to describe shapes of cells.

- *Nanotechnology and nanounits* are being developed to measure and describe the intricacies of the body and perhaps develop technology to combat viruses.

- *Mathematical modeling of breathing* is being utilized to develop new technology to aid and simulate breathing.

- *Adam & Eve 3-D human body maps* are computerized atlases of the male and female human bodies.

- Computers are being used to analyze and identify DNA in a matter of minutes. With such information, cancer cells types are identified and treated with drugs specifically designed to target those specific cells.

To this list must be added one of the most novel approaches being explored to study the workings of the human brain. Who would imagine that geometry would help explain the brain? Mathematicians would! Building on some early work and ideas by mathematicians and statisticians[1], Japanese mathematician Shun'ichi Amari is seeking to unravel mysteries of the brain. He considers human thought to be a composition of logic and intuition. Intuition involves the entire body communicating within itself through an intricate and complex network of neurons. Decisions, thoughts, ideas are created by a combination of logic and intuition. This information is processed and transmitted throughout the body via neurons. Amari is seeking to use mathematical techniques to understand and explain how this process works. What type of mathematics? — a geometry which he has coined and named *informational geometry.* A geometry which will deal with how information is processed and communicated by the brain. In addition, it embodies the mathematics of informational topology, principles of differential

[1] This includes —(1945) Indian statistical logician C. Radhakrishna Rao (who put statistics into a curved geometric structure) — (1970) Soviet statistician Nikolai N. Chentsov — (1979) American statistician Bradley Efron (who pointed out the need to develop a geometry of statistics).

geometry, convergence theory, probability and statistics. As Amari states, "It's a pretty wild idea to use differential geometry to investigate neural networks...We're asking questions like why do information processing systems such as neural networks work so well? What is it that makes them so superior (to traditional computers)? These are questions that require mathematical insights." [2]

The examples linking mathematics and the human body are ever expanding. Consequently, today's medical scientists need to collaborate with mathematicians to learn how to use high powered mathematics to explain the functions of the human body.

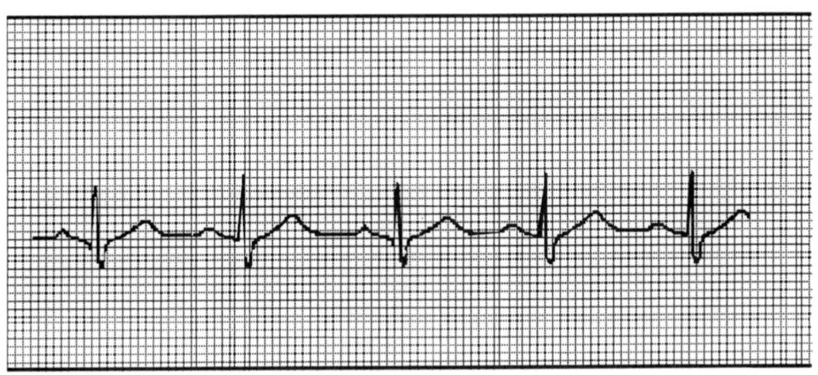

*How a normal heart sinus rthym is graphed.*

---

[2] *Think About It*, Akio Etori. LOOK JAPAN, Feb. 1994.

Its (computer's) destiny is to disappear into our lives like all of the things that we don't think of as technology like wrist watches, paper and pencils.

—Alan Kay

# will computers take the quantum leap?

For the layperson the speed and accuracy of computers seems astonishingly fast and precise. Adding columns of numbers on a spreadsheet, computing one's personal income taxes — these everyday tasks are tedious for us, but child's play for the microcomputer. Yet, for scientist's, the time it takes even supercomputers to perform specific tasks, experiments and calculations may seem sluggish, taking hours, even days to compute and shift through volumes of information. Granted, computer scientists have been ingenious at designing and redesigning new programs

which take short cuts and shave off time from cumbersome computation tasks, but the sequential step-by-step approach required by present day computers can be improved only up to a point. The use of huge numbers and enormous amounts of data — including the search for gigantic prime numbers, factoring methods and cryptography — create the need and possibility of quantum computers.

*What is a quantum computer?* Unlike a conventional computer, a quantum computer would utilize the principles and properties of quantum mechanics. It would be able to take advantage of the multiple energy states that are discussed in quantum theory, as opposed to traditional computers which are designed around the two energy states of electricity, namely "off " and "on". Until recently there was little interest in expending resources to explore this new breed of computer, since conventional computers were successfully being improved upon with new chips, hardware and software. But envision a computer able to attack an enormous number of computational tasks or to follow multiple paths of approach simultaneously. Such a computer would approach certain problems from multiple perspectives. The information from each of the completed tasks would be simultaneously coordinated again and analyzed, and the problem would be solved at breakneck speed. The way such a computer operates would be somewhat analogous to the different ways a novice and a more experienced person would handle the arithmetic problem — $\frac{8}{15} \times \frac{9}{20} \times \frac{5}{2} \times \frac{5}{3}$

The novice would attack the task by actually multiplying the first two fractions, then take that result and multiply it by the third fraction, then that product times the fourth fraction, and finally reduce the result. The experienced person would immediately begin cancelling out common factors throughout the expression, thereby eliminating certain multiplication and reduction steps entirely — arriving at the answer "1" very quickly.

Working intensive problems, such as factoring huge numbers, are the type of tasks that would be simple and well suited for the quantum computer because it would be able to assume multiple energy levels taking many different paths simultaneously in order to solve a problem. In fact, Peter W. Shor of AT&T Bell Laboratories in New Jersey *proved* that the factoring of enormous numbers (numbers with over 100 digits) would be incredibly fast on a computer which functioned along the principles of quantum mechanics. His work provides both motivation for the development of quantum computers and steers quantum mechanics and physics in the direction of computer science. Because of the novel way quantum computers would function, their development and use could be a boon to developing simulations of various scenarios of quantum physics. A quantum computer, by nature, would search out all possiblitlies simulatneously, but once the answer was found all other possiblilties would cease to exist. This charcteristic makes it situable for certain types of problems such as those found in cryptography and the search for the prime factors of enormous numbers. To utilize the special characteristics of quantum computers, new approaches to problem solving would have to be devised. When will computers take the quantum leap? As technologies, such as nuclear magnetic resonance imaging (MRIs) and silicon chips evolve, an electron's spin can be used, explored and followed, and its only a matter of time before tha quatum computer makes its debut.

***What is a quantum?*** An electron can be viewed as a shimmering cloud orbiting the nucleus of an atom. When an electron is forced to change to a smaller orbit it leaps and a quantum of energy (a particle of light) is emitted. If the electron is forced to jump to a larger orbit, then a quantum of energy is absorbed by the atom. As quanta are absorbed and/or radiated, different energy levels are created. These make up the multiple states that a particle assumes. Equally fascinating are the paradoxical properties that these particles possess. For example, how such a particle can be at a variety of places at the same time, enabling it to take all possible paths simultaneously is still a mystery. Once a state is observed, the other possibilities cease to exist. Similarly, since quantum computers will use *qubits* for their bits (a qubit can be a 0, a 1, or a mixture of 0s and 1s indicating the various electron spins), once a qubit is measured it becomes a 0 or 1.

Mounatins are not pyramids and trees are not cones. God must love gunnery and arhictecture if Euclid is his only geometer.

—Tom Stoppard
*Arcadia*

# mathematics, Guggenheim Bilbao and Frank Gehry

Architecture, like art, is innately tied to the ideas of mathematics and physics. Without adhering to their concepts, the construction of a building and its integrity are in question. When we look around us the objects from Euclidean geometry jump out first. We see shapes which have been ingrained in us from our early years such as circles, spheres, squares, triangles, rectangles, solids; shapes which resemble some of nature's objects such as the disc of the moon, the

*Riverfront view of the museum. Riverfront promenade and water garden. Photograph by David Heald. Courtesy of the Guggenheim Museum Bilbao. ©SRGH. New York. 1997.*

shape of an orange, the formations of crystals. Looking more closely and from a non-Euclidean perspective, we see shapes that were never described by Euclid which are a dominant part in the things around us. Most architectural designs, especially before the 20th century, relied heavily on Euclidean forms as models. Today, designs have broken away from that mold, and embrace some of the forms and ideas of non-Euclidean geometry. New materials, building techniques and technologies allow the architect to realize structures hitherto only imagined. St. Mary's Cathedral in San Francisco, California, built in 1970, with its massive hyperbolic paraboloid roof, is such a design. The Guggenheim Museum Bilbao, designed by Frank Gehry, is yet another. It ushers in the architecture of the 21st century with its structures speaking boldly of new architectural shapes and forms that until recently could have existed only in the architect's imagination.

*Photograph by David Heald. Courtesy of the Guggenheim Museum Bilbao. ©SRGH. New York. 1997.*

This awesome building illustrates the transformation of Euclidean solids by means of topology to non-traditional forms. In Gehry's creation we see *rubber sheet geometry* at work. One can almost sense the pulling and stretching of shapes that result in this masterpiece.

Gehry's creation is exquisitely in harmony with its triangular shaped site located along the Nervion River in Bilbao, Spain. One is struck by the commanding *presence* of the Guggenheim Museum Bilbao. Its shapes immediately capture our attention. In fact, it is difficult not to stare at the series of interconnecting shapes, vested in limestone, crowned with a titanium metallic roof (called the *Metallic Flower* ) which is graced with phenomenal complex curves. It is a feast for the eyes!

Advanced computer technologies have played an indispensable role in the realization of this structure. Using *Catia,* an advanced 3-dimensional modeler developed to map curved surfaces for the aerospace industry, Gehry was able to explore and alter shapes while remaining within the parameters of feasibility of constructible geometric forms with specific materials being utilized. Every point of the model's surface was mapped by using a digitizing process that traced the model's shape with a special arm-like digitizing tool. The information, when transferred to *Catia,* allowed the architect to explore shapes and maintain the geometric relationships to the constructibility of the model's shapes. While controlling a milling machine, *Catia* carved an exact model of the building's forms, maintaining dimensional control for the building's systems for the various materials used. Gehry points out that this new technology "provides a way for me to get closer to the craft. In the past, there were many layers between my rough sketch and the final building, and the feeling of the design could get lost before it reached the craftsman. It feels like I've been speaking a foreign language, and now, all of a sudden, the craftsman understands me. In this case, the computer is not dehumanizing; it's an interpreter."

The museum's complex exterior deceptively conceals a relatively simple and flowing interior. In addition to housing traditionally sized galleries, its design includes unique large-scale presentation spaces (the largest is 450 feet by 80 feet) whose space remains unobstructed by structural columns. Besides the three levels of gallery spaces, the museum has an

*Photograph by David Heald. Courtesy of the Guggenheim Museum Bilbao. ©SRGH. New York. 1997.*

auditorium, restaurant, and cafe all of which encompass a 165 foot high atrium bathed with natural light from the *Metallic Flower* tower.

The Guggenheim Museum Bilbao is much more than a museum. It is a work melding modern art, technology and mathematics.

...a theorem as 'the square of the hypotenuse of a right angled triangle is equal to the sum of the squares of the sides' is as dazzlingly beautiful now as it was n the day when Pythagoras discovered it.
—Charles Dodgson
(Lewis Carroll)

# the Pythagorean theorem

## — *the survivor*

The Pythagorean theorem's footprint surfaced and resurfaced over the centuries in many places and was used by many different civilizations. Many of us have heard of it. Some have struggled with its proof in geometry class. And a few of us may have created our own proof of it. When exactly this mathematical jewel was developed is unknown. The mathematical notion that a right triangle's sides always are related by a formula and that certain values for the sides of a triangle will always produce a right triangle was discovered thousands of years ago — probably around the time of the Babylonians. We know that the Babylonians were familiar with the theorem and with Pythagorean triplets[1] because they recorded Pythagorean problems and tables of Pythagorean triplets on their cuneiform tablets[2]. The Egyptian rope stretchers utilized ropes knotted with a Pythagorean triplet to create a right angle for their structures. Although the notion of the theorem had been around the ancient world for centuries[3] prior to its proof, this notion had not been taken one step further — abstracted, generalized and proven. This was probably first done by someone from the Pythagorean school around 540 BC, hence its name.

But the Pythagorean theorem story did not end with its proof. There is probably no other theorem that has been proven more often in so many ways by so many people from different continents, cultures and centuries. Among these we find Euclid's *The Forty-Seventh Proposition* to be the earliest existing proof. Euclid also developed a proof of the converse of the theorem. Other well known proofs of this theorem were done by:

- Oliver Byrne (1810-1880), which was a multicolored version of Euclid's proof.
- President James Garfield (1831-1881).
- Hindu scholar Bhãskara circa1150, which proof appears in his book *Vijaganita*, consisting mainly of sequential diagrams.
- Leonardo da Vinci (1452-1519).
- an unknown Chinese scholar, which appeared in the manuscript *The Chou Pei*, whose date is disputed to be bewteen 1200 B.C. and 100 A.D.)

---

[1] A Pythagprean triplet is a set of 3 positive numbers, a,b, and c, which satisfy the equation $a^2+b^2=c^2$. For example, 5, 12, and 13 work, and a triangle with sides the length of these three numbers is a right triangle.

[2] *Science Awakening* (by B.L. van der Waerden)

[3] Evidence of this theorem dates back to the Babylonians of Hammurabi's time, over 1000 years before Pythagoras.

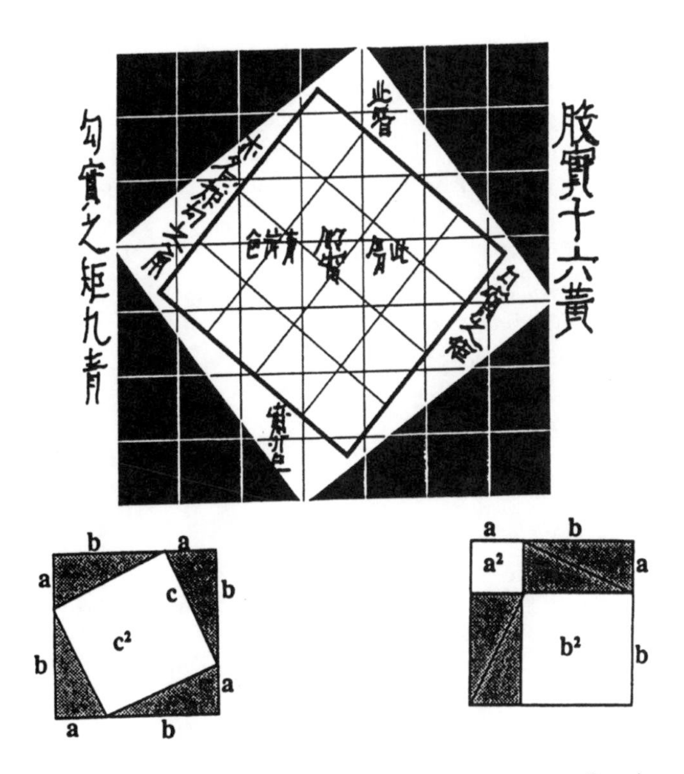


*The diagrams above, which illustrate the Pythagorean theorem, are from the Chinese manuscript **The Chou Pei** (with conflicting dates anywhere from 1200 B.C. to 100 A.D.). When these are studied and rearranged, the proof of the Pythagorean theorem relationship, $a^2+b^2=c^2$, materializes.*

The theorem also gave rise to the following variations:

*—The Greek mathematician Pappus proved the theorem which dealt with areas of parallelograms built on the sides of a right triangle.*

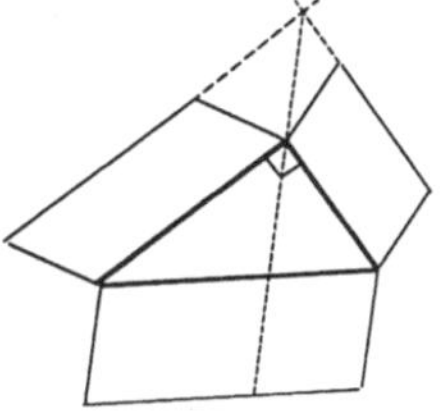

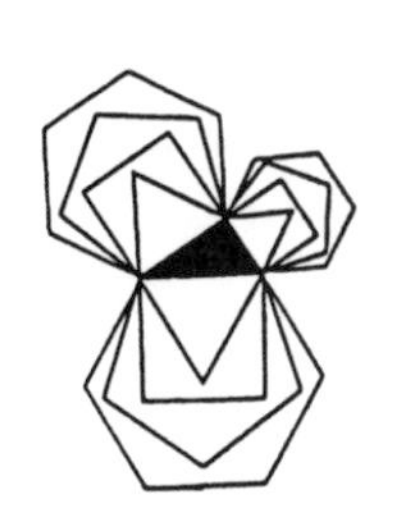

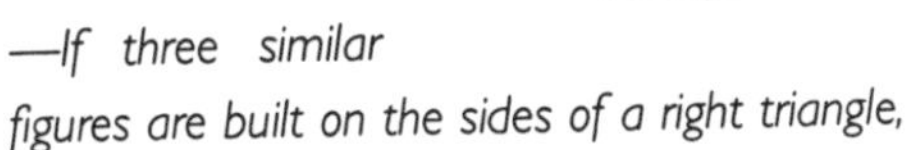

*—If three similar figures are built on the sides of a right triangle,*

*then the sum of the areas of the two built on the legs equals the area of the one built on the hypotenuse.*

The usefulness and validity of this mathematical idea has transcended centuries and left its mark on many fields of mathematics. Notice that:

- Though the theorem was originally conceptualized on a flat plane, the theorem is not restricted to two dimensions. It provides the means for determining the distance between two points in a plane, in space, in hyperspace, in fact, in n-dimensional space.

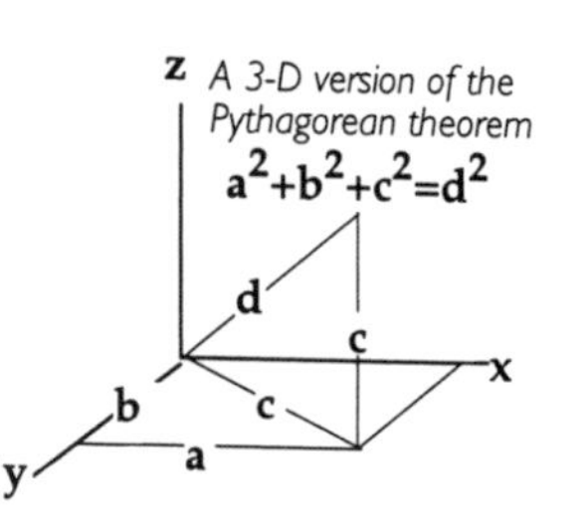

*A 3-D version of the Pythagorean theorem*

- It is used in the proof of the quadratic formula.

- It is hidden in the well known trigonometric identity, $\sin^2+\cos^2=1$.

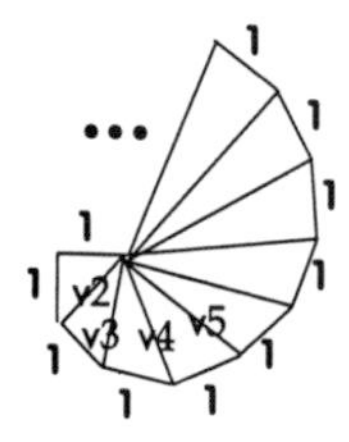


- It gave rise to irrational numbers.[4]

- It even has applications in non-Euclidean geometry!

It is a theorem whose usefulness and importance goes on and on!

---

[4] When the diagonal of a square with sides 1 unit is drawn, the Pythagorean theorem establishes the diagonal's length as the v2. It is ironic that the Pythagorean theorem is responsible for the demise of the Pythagorean's mystique of the whole numbers as rulers of the universe. In fact, legend has it that the Pythagoreans wanted to keep the irrationals secret.

What is nature?
Nothing in relation to the infinite, all in relation to nothing, a mean between nothing and everything.

—Blaise Pascal

# rings, helices & dolphins

## *mathematics & nature*

Ancient tales in many cultures characterize dolphins as benevolent creatures, often saving the lives of humans by guiding them to safety. Today, we know about their special sonar abilities, and efforts are being made to establish methods of communication. Other scientific research is being conducted to assess the self-awareness and intelligence of these magnificent creatures. Globally, dolphins, both in the wild and in captivity, have been observed toying with bubbles. But it is more than just playing with these objects. Many animals have toys, but what crea-

*Photograph courtesy of Project Delphis at Earthtrus. Honolulu, Hawaii.*

tures other than humans actually try to develop and improve their recreational skills and pastimes? We see the sailboarder repeatedly dumped before understanding the mechanics and physics of the board, the sail and the wind. We find the consummate basketball player practicing free throws, and the computer game addict glued to the computer monitor. Such an approach to activities has only been associated with humans, yet global research on dolphin intelligence has revealed some fascinating findings. These observations have not been limited to a specific dolphin, but have been seen among the Amazon river dolphins, bottlenose dolphins, and Pacific and Atlantic spotted dolphins. In mathematics we think of the sphere, the torus, the spiral and the helix as sophisticated math shapes and not necessarily as toys. But apparently to dolphins they are a form of entertainment.

How do dolphins create these toys and use them? Studying bottlenose dolphins in an underwater laboratory at Sea Life Park in Honolulu, Project Delphis[1] conducts scientific research on dolphin intelligence and behavior. Here a group of dolphins has displayed a special affinity and interest in *bubble math.* In topology a sphere cannot be transformed into a ring (or torus). No matter how it is stretched, shrunk, distorted it never becomes equivalent to a torus. In the world of dolphins and the laws of physics of water, however, a sphere has no trouble being transformed into a ring. What is so special about playing with bubble rings? Dolphins do not make these rings or other bubble formations on command, nor do they do it for reward, food or sexual activity. They have been observed to create bubbles for recreation and to enhance their communication. It is believed that a dolphin receiving both sonar and bubble signals uses its own sonar to help decipher the intended message. In addition, observations at Project Delphis have revealed how certain dolphins have mastered the physics of underwater bubbles and the use of vortices. Since the water pressure under the bubble is greater[2] than above it, this pressure both lifts the bubble vertically and pushes a hole through the center of the sphere[3], thereby forming a ring.

---

[1] Ken Marten is director of research for Project Delphis. Under the auspices of Earthtrust, the project was initiated by Don J. White in 1985 to investigate and assess dolphin intelligence and as a conservation effort to save wild dolphins. White founded Earthtrust in 1976 with the profound hope that new perceptions of dolphin intelligence would motivate humans to respect and protect dolphins and their natural habitat. For additional information and photographs contact World Wide Web site at http://earthtrust.org. and e-mail: earthtrust@aloha.net.

[2] Water pressure increases the further below the surface a point is.

[3] This happens as long as the sphere's diameter is at least 2cm.

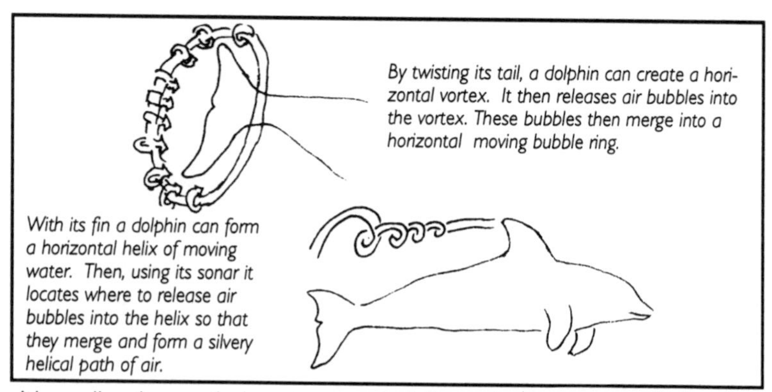


Normally, the vertical vortex created through the center of the ring carries the bubble straight to the water's surface. But these dolphins have figured out how vortices work, and have developed means to have rings move horizontally rather than vertically. Furthermore, by experimenting, certain dolphins have learned to use their fins and body movement to create horizontal vortices. Utilizing their sonar they find the exact location of the eye or center of the vortex. Here the pressure in the vortex is lowest, and it is here the dolphin releases air bubbles that follow the helical path of the vortex and merge together to form a helical air path that moves horizontally. Dolphins that have perfected their skills produce smooth and shiny bubble sculptures which stand out as works of art when compared to those done by novice dolphins. Another intriguing factor is that the dolphins' bubble play and experimentation are not something all dolphins do naturally. In fact, one group of dolphins in a Project Delphis study were not making bubble rings until a dolphin who knew how to make bubbles was introduced into their group. Intrigued by this newcomer's rings, other dolphins in the tank decided to learn the "game". In addition, dolphins have been observed teaching one another how to create the various bubble formations. Bubble sculptures are not a fluke, but take time, patience and ingenuity to master and imagination to create new forms reminiscent of a kind of kinetic art and science.

# the art of Claude Monet

Mathematical genius and artistic genius touch one another.

—Gösta Mittag-Leffler

Some artists' works call to mind mathematics because they were conceived with mathematics in mind. Among these we find — M.C. Escher's *Metamorphosis*, focusing on the use of tessellations; Leonardo da Vinci's horses, relying on mathematics of proportions; Isamu Noguchi's *Red Cube*, using mathematical objects as its model. Many artists' works are and were influenced by the mathematical ideas and technology of their time. Renaissance artists relied heavily on mathematics, incorporating ideas from projective geometry to add realism and depth. They even

*The famous Japanese bridge as it is today at Giverny.*

developed tools to enhance their ability to capture the proper perspective by using such instruments as da Vinci's spectograph and Albrecht Dürer's projecting scene apparatus. Artists such as Pablo Picasso and Marcel Duchamp used the underlying theme of multi-dimensions and time in works of cubism. Today, computers are used extensively by many artists in their designs— as in the sculptures of Bruce Breasley, Ronald Dale Resch, Tony Robbin, Helaman Ferguson, and the designs of Mitsumasa Anno.

The majority of artists' works were conceived without considering mathematics. Mathematics could not have been further from their minds. But *we* can analyze, look for and discover mathematical ideas present in much of their work. Among these artists is Claude Monet. In his lovely scenes of Giverny, with scene after scene of the Japanese bridge, we are enthralled by the his use of colors and light and the blending of strokes

*A closeup of a group of Monet's waterlilies.*

making the water lilies become distinct as we move away from the painting. Mathematics is far from our minds as we enjoy his works. In retrospect, however, we can uncover mathematical ideas in his revolutionary paintings. He was one of the first to incorporate the element of time in his art. Besides having the parameters of the 3-dimensions of Euclidean geometry, Monet added the dimension of time by painting many of his works as a series — a series governed by time. Sometimes he painted the same scene over and over during different seasons, as with the Japanese bridge at Giverny. Other times the series may have followed a smaller interval of time — a day. He studied how light changes a scene, as in the series of the *Rouen*

*A closeup of one of Monet's famous waterlilies.*

*Catheral.* These works can be considered the elements of a sequence. A sequence governed by time. What Monet saw determined the pattern of the sequence —the small incidental changes caused by the passage of time. Monet captured stages of an evolving scene, even as today mathematics and the computer capture the evolving stages of a fractal.

Monet is famous for his phenomenal impressionistic work. Again, putting on mathematical glasses, the elements of impressionistic art can be given a new face, a *fuzzy* one. In Monet's *Waterlilies and Japanese Bridge* or in *Garden at Giverny* there are fuzzy objects. His waterlilies, for example, when viewed up close do not look exactly like lilies. In other words, these lilies cannot be regarded 100% water lilies 100% of the time because it depends upon how they are viewed — by whom and from which vantage point. Thus, the mathematical adjective *fuzzy* — having nothing to do with focus, but rather identity as in the eye of the beholder ( i.e. it is relative to the viewer) — has given us a new way to describe impressionistic works.

Are these mathematical ideas applied to Monet farfetched? Why not look for mathematics here even as we look for it to describe other phenomena around us?

# how knots are tied to mathematics

But the creative principle resides in mathematics. In a certain sense, therefore, I hold true that pure thought can grasp reality, as the ancients dreamed.

—Albert Einstein

One of the earliest connections between mathematics and knots was the use of knots to signify numerals. Such knots appeared on lo-shu, the ancient Chinese magic square, where the black knots signified the even numerals and the white the odd numerals. A knot was also the form given to the ancient puzzle, the Gordian knot, most often associated with Alexander the Great. Other examples where knots tied on strings or cords were linked to mathematics include:

- *to keep track of days —*

*From antiquity knots have been used as decorative designs. These designs pique our interest to follow the turns of a knot as one does the paths of a maze. This Celtic design was painted by monks in Northumberland circa 700.*

When he left on an expedition, King Darius of Persia left a thong tied with 60 knots for his court to keep track of the days before his return.

2 hundreds
6 tens
7 ones

• *to record wages* — Workmen of the Ryukyu Islands devised knot-type placevalue system with the fringes of braided reeds.

• *to count prayers* — By using knots, rosaries and Tibetan prayer strings are still used to keep track of the number of prayers.

• *to label products* —In the early 1900s German millers from the Baden province in labeled sacks of flour with knots in the cords cinching the sacks. The type of flour and meal was indicated by loops while the quantity was shown with specific

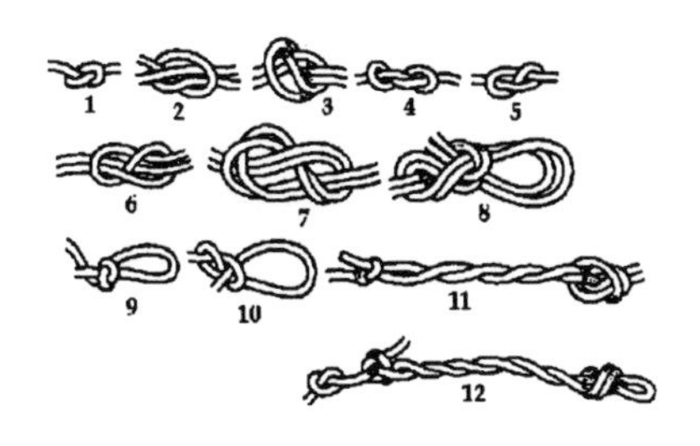

knots.

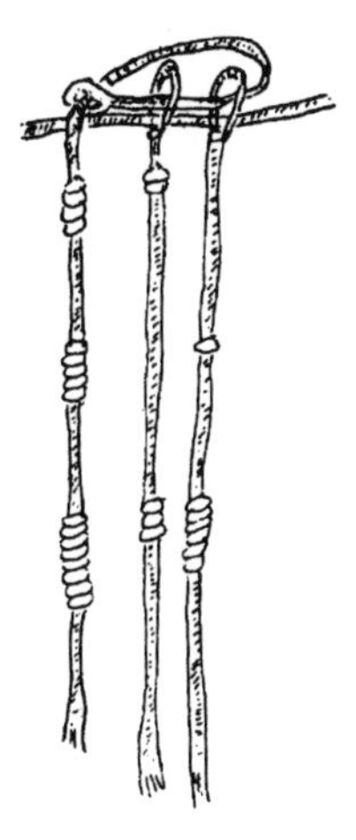

• *to store data* —The Incas recorded information about crop yields, taxes and population on quipus (knotted ropes) that used a base ten placevalue system.

*One was signified by a single knot, 2 a figure eight knot, 3 had three loops on a type of slip knot; 2 to 9 had two to nine loops on a slip knot; the first knot near the bottom of the cord was the ones place, the next row up was the tens, the next above that was the hundreds, a space was zero. So 458 appears in the first cord; 230 in the second cord; 62 is on the third cord; the cord holding the three cords represents the sum of these cords, namely 750. The scribes also used knots of quipus to record non-numerical information somewhat analogous to how binary numbers record word processing on computers or how symbols are used by court recorders.*

In the last hundred and fifty years, the mathematical connection of knots has taken a different twist. Today, mathematicians are developing the mathematics of knots, called *knot theory*, a branch of topology that studies the actual structure of knots. These mathematical knots are different. They have no loose ends. For example, this knot is not classified as a knot in knot theory. But if you join the loose

ends, then it is known as the *trefoil knot.* When one takes the traditional figure-8 knot and joins the loose ends, it becomes the four-knot of topology.

*What do mathematicians do with such knots?* First, they study their characteristics, patterns, and try to classify them. In topology a knot is defined as a closed loop in three-dimensional space. In topology an object's size, shape, and location in space have no significance. In addition, the tautness of a knot is unimportant. Knots are classified and characterized by identifying properties that do not change (called invariants) as the knot (a topological object) is moved around (transformed) in various ways. For example, one property called the knot's *crossing number,* is assigned to a knot by counting the number of times the knot's strand crosses itself when it is in its simplest knot form (with no superfluous twists). Since this number is not distinct for each knot, other ways of distinguishing knots are needed. In fact, algebraic polynomials (Alexander polynomials of 1928, Jones polynomials[1] of 1984) have been devised to also describe knots[2]. Among other knotty ideas, mathematicians have defined: ways to add knots, the null knot, prime knots. In other words, knots have literally become mathematical objects.

*How are these knots used?* As with so many mathematical ideas, their uses are an ever evolving process. Today, knot theory is being used in:

- *genetics* — Biologists and mathematicians have joined forces to understand what type of knotting allows DNA to replicate so easily.

---

[1] Named after mathematicians who developed these polynomials, J.W. Alexander and Vaughan Jones. In 1990 in Moscow, mathematician Victor Vassiliev developed a method of computing numerical knot invariants, called *graphs.*, which have been subsequently found to be connected to knot polynomials. These graphs provide new insights on knot classification.

[2] $x^2-x+1$ is the Alexander polynomial for trefoil knot.

- *molecular level* — The behavior of liquids and gases in conjunction with statistical mechanics is studied.

- *physics* and in particular *topological quantum field theory* — Knot configurations can be used to describe the different interactions of particles that can take place. In addition, physicists are exploring *superstrings*, ultra-small knots encased in higher dimensions that may hold answers to the make-up of matter and energy of the universe.

- *chemistry* — Knots are used to distinguish chemical elements by visualizing their atoms as distinct knotted vortex tubes.

The more knots are examined with a mathematical eye, the more we are finding their ties to many aspects of our lives and the universe. Could it be that just as humans over the centuries utilized knots in numeration, accounting, art, pastimes, and for protection that nature also relies on knots in its creations.?

What we know is not much. What we do not know is immense.

—Pierre-Simon de Laplace

# the Chinese remainder theorem

Problems and their solutions are at the crux of mathematics. They kindle ideas and tax minds. Dating back to earliest ancient records we find —Egyptian problems in the Rind papyrus, Babylonian problems written on clay tablets, the three famous problems of antiquity and many others by Greeks such as recreation problems attributed to Metrodorus (circa 500BC), a host of problems from India, China and Japan, a Medieval collection of Alcuin of York *(Propositiones ad acuendos juvenes,* which included recreation problems for Charlemagne), the treasure trove of problems from the Middle ages dealing with chess boards, inheritances, pursuit, trade, etc.. The list goes on and on. Most of these problems, in various formats, have found their way to our present day text books, and many have plagued mathematicians for centuries, as witness the proof of *Fermat's Last Theorem.*

One fascinating problem from the East is—

> ***Find a number so that when it is divided by 3 the remainder is 1, when it is divided by 5 the remainder is 4, and when it is divided by 7 the remainder is 2.***

A remainder problem similar to this originally appeared in the work of Chinese mathematician and philosopher Sun Tsu. The date of his writings, *Sun Tzu Suan-ching* was around 200 A.D.. Leonardo da Pisa (c.117-1250), more frequently referred to as Fibonacci, also presented such a problem in his book *Liber Abaci.* Today, such problems are classified under the Chinese Remainder Theorem and appear in the field of mathematics known as number theory.

| **remainders for, divisors** | **3** | **5** | **7** |
|---|---|---|---|
| **numbers divided** | | | |
| **0** | 0 | 0 | 0 |
| **1** | 1 | 1 | 1 |
| **2** | 2 | 2 | 2 |
| **3** | 0 | 3 | 3 |
| **4** | 1 | 4 | 4 |
| **5** | 2 | 0 | 5 |
| **6** | 0 | 1 | 6 |
| **7** | 1 | 2 | 0 |
| **8** | 2 | 3 | 1 |
| **9** | 0 | 4 | 2 |
| **10** | 1 | 0 | 3 |
| **11** | 2 | 1 | 4 |
| **12** | 0 | 2 | 5 |
| **13** | 1 | 3 | 6 |

Fibonacci originally introduced this problem as a parlor game, in which a set of divisors did not necessarily change but the remainders changed during the game. The divisors can vary and you can use as many as you want. But the divisors must be relatively prime to one another.

*So what is the strategy of this*

不知其數三三數之賸二
答曰二十三
曰三三數之賸二置一百
賸三置六十三七七數之
得二百三十三以二百一
三三數之賸一則置七十
則置二十一七七數之
百六以上以一百五減之

*A portion of a table illustrating a remainder problem written in Chinese script numerals*

*problem, and what do the columns of Chinese script numerals reveal?*

The table above shows the remainders when the numbers in the first column are divided by the respective divisors in the top row. Looking closely we notice no number has the same set of remainders, because we used divisors that were relatively prime[1] to one another. There will be no repetitions for the first 105 numbers ($3 \cdot 5 \cdot 7 = 105$). With the number 106 the table repeats. This cyclical repetition was used by Sun Tsu in his astronomical work.

---

[1] They have no divisor in common other than 1 or -1, for example 5 and 12 are relatively prime to one another.

***How to solve the problem?***

Recall the remainders in the problem presented were:

*1* when divided by **3**, *4* when divided by **5**, *2* when divided by **7**. We are looking for a number, call it **n,** that has these remainders when divided independently by 3, 5 and 7. Let's consider another number, call it

m = **1**(70) + **4**(21) +**2**(15).

Where did 70, 21 and 15 come from? For the divisor **3** we take the product of the other other two divisors, namely 5•7=35, and find the first multiple of this number that is 1 unit greater than a multiple of 3. Since 35 is not 1 unit greater than 3's closest multiple, 33, we try the next multiple of 35, namely *70* which happens to be 1 unit greater than 69, which is a multiple of 3. For **5** we take 3•7=*21*, which is 1 unit larger than a multiple of 5 ( i.e. 20). For **7** we take 3•5=*15*, which also is 1 unit larger than a multiple of 7 (i.e. 14). So *70, 21* and *15* are the special multipliers for the divisors **3, 5** and **7**. Each set of divisors has its special set of multipliers, and these are the ones for **3, 5,** and **7**. When you have these multipliers, you have essentially solved the problem because all you need to do is compute the following:

m = 184 and find its remainder
after you divide it by 105 (=3•5•7)
which is **79**, our solution.

You can use these multipliers for any remainders when dividing by **3** , **5** or **7**.

* * *

Concepts from number theory explain why this procedure works. When **m** is divided by **3** its remainder is 1 because of the way we selected the multipliers 70, 21, and 15. **3** goes evenly into **4•21** and **2•15** but not into **1•70**, for which it has a remainder of *1*. Similarly for **5.** **5** goes evenly into **1•70** and into **2•15** but leaves a remainder of 4 for **4•21**. For **7** we get the remainder 2.

By examining an interesting property in number theory, we shed more light on how this process works. For example, we know that 23 divided by 4 has a remainder of 3, and 7 divided by 4 also has a remainder of 3. But (23-7) divided by 4 has **no** remainder because in the subtraction process the remainders eliminated one another since they were the same remainder for each. So in our problem since **m** and **n** both have the same remainders when divide by 3, 5 and 7, then **m-n** has *no* remainder when they are divided by $3 \cdot 5 \cdot 7$ (=105). Thus,

$$\begin{aligned} m/105 &= (1 \cdot 70 + 4 \cdot 21 + 2 \cdot 15)/105 \\ &= 184/105 \\ &= 1\ 79/105, \end{aligned}$$

and the remainder **79** is the number **n** we are looking for.

There is no question that mathematical problems will always tease, tantalize and intrigue our minds.

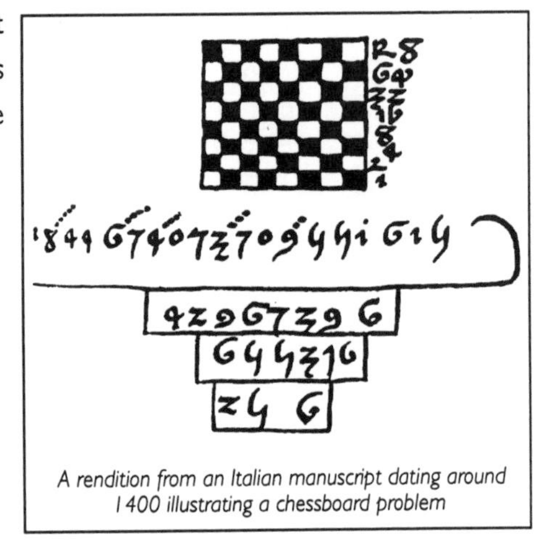

*A rendition from an Italian manuscript dating around 1400 illustrating a chessboard problem*

Although to penetrate into the intimate mysteries of nature and hence to learn the true causes of phenomena is not allowed to us, nevertheless it can happen that a certain fictive hypothesis may suffice for explaining many phenomena.

—Leonhard Euler

# solitons

Ever become mesmerized by watching waves at the beach — staring at the rhythm of the waves, the sets, trying to see patterns, trying to predict a big one? Imagine a never changing single wave. A wave traveling across great distances of its medium, yet never changing shape; just constantly moving the same way. Impossible, you say. Many believed that such waves were not real, but phantom waves. Meet a *soliton* — the wave thought to be a figment of the imagination. Scientists are finding soliton waves to be more frequent in nature than ever imagined.

When such a wave was first recorded by engineer John Scott

Russell in 1834, scientists refuted his observations as impossible. Witnessing the slippage of a barge creating an unusual single wave action, the young engineer watched as the wave formed and moved down a canal. Captivated, Russell followed the wave as far as he could on horseback. He recorded that it never changed shape or speed. Working in his laboratory, he observed, the taller a wave was the faster it moved.

For years scientists had rejected the existence of such waves. It was not until 1895 that mathematicians Diederik Johannes Korteweg and Hendrik de Vries realized that opposing forces (dispersion and compression), which normally dissipate a wave, could actually reach an equilibrium so that the wave would remain constant. But they felt such natural occurrences were very rare. By the late 1800s non-linear equations had been adapted to describe solitons.

In 1965 mathematicians Martin Kruskal of Princeton and Norman Zabusky of Bell Laboratories observed how these waves remained unchanged when two waves of varying speeds happened to meet. They would just pass through one another without effecting their individual heights or speeds. The only change observed was a slight variation in the phase shift (the faster one ended up a bit ahead of where it would have been and the slower a bit behind). Thus, they named them *solitons*. Solitons are actually common in nature. In fact, they could exist in such diverse mediums as water, air, earth, electromagnetic fields. Solitons are complex systems whose forces amazingly are kept in perfect balance so the wave does not crest and does not dissipate. Such a wave is constantly being adjusted by contradicting forces — each cancelling the other out so that no single force overcomes and changes the wave's movement.

***Where does one find natural occurring solitons?***

> In *genetics*, solitons are being explored in the way the DNA double helix molecule replicates itself — separating or "unzipping" of the double helix so that free nucleotides can

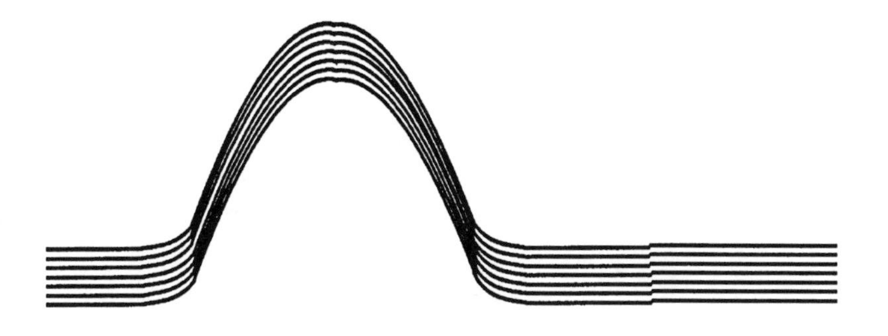

attach to the two strands of the split helix and form two new identical DNA double helices. This unzipping and zipping action may take place in the form of a soliton wave!

In *molecular biology*, soliton action[1] is being considered to explain the role proteins play in the movement of muscles. A soliton wave could be an efficient way to move a glob of energy. In the brain, solitons may be the means by which sodium atoms (charged particles) outside the brain cells move by riding the soliton wave through the proteins.

---

[1] A soliton would be a "very efficient way to concentrate energy and get it to the right place," says Alwyn Scott of the University of Arizona. See *Lone Wave* by David H. Freedman. *Discover* magazine, December 1994.

In *cosmology*, solitons again come into the picture. Some scientists are considering soliton stars as alternatives to black holes. Hong-Yee Chiun of the Goddard Space Flight Center proposes a scenario in which after the Big Bang quarks were trapped and squeezed in a collapsing pocket of high-energy space, thereby forming a soliton star. This soliton star is capable of pulling in surrounding matter whose protons and neutrons are changed to free quarks, thereby allowing the soliton star to grow even larger. Others contend that since soliton stars can be so enormously massive and invisible, perhaps they are what comprise the dark matter that scientists are trying to explain.

In *quantum theory* solitons are being considered as what particles may look like, since solitons may possess properties of both a wave and a particle.

Solitons are not only being explored in nature, but in *industrial applications* as well. Because solitons retain their identities when they encounter one another, these waves are being considered, for example, in light propagation in optical fibers and in medical lasers to produce a soliton energy wave at a specified distance, thereby avoiding the need to pierce healthy tissues. Allan Snyder and colleagues at Australian National University in Canberra are exploring the use of solitons in optical fibers by sending bits of light over long distances packed into a soliton wave. Scientists at Corning Inc. have designed the fiber so that the solitons can retain their original shape over long distances. At the University of Rochester, Andrew Stentz demonstrated that these impulses can travel along the fiber as fast as one-trillionth of second without degrading, which would be 100 times faster than current methods. Mathematicians Philip Rosenau (of the Techion in Haifa, Israel) and Mac Hyman (of the Theoretical

Division at Los Alamos) are exploring what they call *compactons* (solitons without tails). These can theoretically be packed full of information and transmitted in a line along the Internet.

Japanese researchers at the University of Osaka are working on perfecting their creations of the first low frequency soliton sound waves. They hope to be able to use these waves in developing new methods to transport heat efficiently through pipes.

Expect to hear more about these "phantom" waves, which are described by non-linear mathematics[2]. They could well pop-up in any familiar wave action —earthquakes, ocean waves, radio waves or exotic waves of quantum physics and genetics.

---

[2] Non-linear mathematics studies complex systems (non-linear systems) which are governed by an enormous number of diverse factors. These factors are in a constant state of changing and adjusting. Complex systems are on the edge of chaos and order in an ever balancing state of equilibrium, which relies on self organizing dynamics of the system. In other words, it is constantly adapting and balancing itself to meet the changing factors and circumstances that are taking place. Non-linear mathematics draws on mathematics from probability, chaos theory, fractals, artificial intelligence, fuzzy logic, computer science and other fields.

# the mathematics of weather forecasting

Everyone talks about the weather, but nobody does anything about it.
— Mark Twain

*"Areas of low clouds and fog clearing by midday; otherwise sunny and warmer. Highs 67-71°F. Low clouds tonight. Lows 47-52°F."*

The weather impacts our lives in so many ways and can affect plans, travel, work, moods. Weather is a complex system affected by temperatures, humidities, barometric pressures, winds, cloud motions, atmospheric conditions, vegetation or lack of, solar conditions, oceans temperatures , ...and an infinite number of minute and varied factors which go undetected, thus

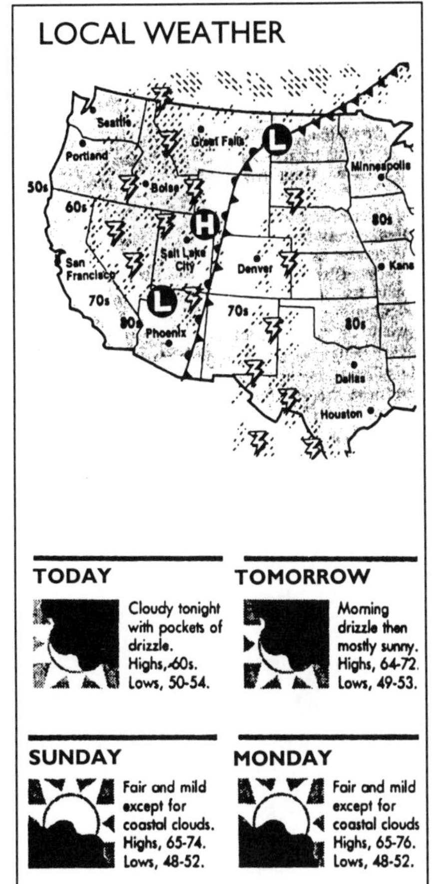


explaining why forecasts are good for only a few days at a time. The factors that come into play are non-linear. They affect the reliability of weather forecasting since a specific set of conditions does not necessarily produce the same weather.[1]

The science/art of weather forecasting has been developing for centuries with ever evolving new techniques and tools. Today, the predictability/unpredictability of weather forecasting depends on measuring and mathematizing vast amounts of meteorological data. Presently, it is physically impossible to detect, register and input all data, variations and changes that affect the weather. Since factors do not react linearly, chaos and complexity play major roles . Undetected slight variations may trigger other minute changes which can make a weather forecast invalid and can create a weather surprise. Daily forecasting has improved immensely from the times predictions had to rely only on data from a limited number of weather balloons and stations. Today, the speed and amount of

[1] For a linear system, a specific input always gives the same output.

In 1648, from atop the Tour de St. Jacques (Paris, France), mathematician Blaise Pascal used a barometer to calculate the weight of air . Today, the tower houses meteorological equipment.

sampling data have increased dramatically. Radar systems and weather satellites[2] coupled with a complex mathematical algorithm have improved the reliability of forecasts.

---

2 These include the various types of GOES (Geostationary Observational Environmental) satellites with VAS (visible infrared spin-scan radiometric atmospheric sounder) instruments).

In 1999 at the University of Wisconsin in Madison, new computer programs were developed with pattern recognition capabilities which actually associate a number with a hurricane's intensity. The program scans satellite images and then quantifies the strength of the storm. In addition, new hurricane tracking techniques were also developed. The images generated from a group of satellites (each with a different view of the earth) are merged into a high resolution graphic from which hurricane information can be extracted.

***How is a forecast created?*** The NWS (National Weather Service) collects information observed from around the globe. This data comes from human observations, weather balloons, planes, boats, stations, and satellites. On a supercomputer the NWS runs a numerical weather prediction algorithm which divides the atmosphere into a 3-D grid with over 250,000 points. The collected data gives values for temperatures, pressure, moisture, and wind velocity for each of these points. The computer produces a model of the atmosphere from the data at the time it was gathered. Using the data gathered, specific equations are applied to predict 10 minutes future values for each point of the grid. The process is repeated until a 24 hour weather model is pictured and a forecast formulated.

Nevertheless, even with the use of supercomputers and the accuracy of complex computer modeling, a forecast's reliability is good for only a matter of days. New methods and improvements are constantly being explored both for short and long term predictions. Presently seasonal forecasts are based primarily on statistics from past weather patterns and past analogous situations thereby creating a forecast whose accuracy is limited to at most three months at the most. One new approach being explored for long range forecasting uses a tool called a *coupled global circulation model* (GCM). GCM simulates how streams in the oceans[3]

---

[3] Ocean streams such as El Niño (an occasional warming in the tropical Pacific) and La Niña (an occasional cooling of tropical Pacific waters) have been found to influence weather.

and atmosphere move heat and moisture around the planet. The preliminary work with GCM is very encouraging[4], and the National Oceanic and Atmospheric Administration set the year 2005 as the target date for being able to forecast general climatic conditions up to a year in advance with a 70% success rate.[5] Long range forecasting is a revolutionary approach to weather reporting, since it does not focus on specific time predictions, but upon general climatic conditions for a year or more in advance. What changes in methods are in the offing for projecting daily forecasts? For improvements in short term predictions, look to the new dynamics of computer modeling coupled with innovations from the science of complexity, in which mathematics plays a key role.

---

[4] Hindsight forecasts are being made some ten years in the past, to test the model's accuracy, and explore ways to modify it.

[5] See *The Long View of Weather* by Richard Monasterky, Science News, Nov. 20, 1993 for additional information on GCM.

*mathematics & the architecture of*

# Arata Isozaki

From the intrinsic evidence of his creations, the Great Architect of the Universe now begins to appear as a pure mathematician.

—James H. Jeans

All the shapes of architecture are based on mathematical objects, be they the geodesic domes of Buckminster Fuller, the golden rectangle of the Parthenon, the pyramids of Egypt, or the hyperbolic paraboloid of St. Mary's cathedral in San Francisco. Mathematical shapes and curves are the building blocks for an architect's design. For Arata Isozaki these shapes are also the symbols of a language he uses to make his creations speak. As he points out, "Each (structure) must create its own independent symbolic self".[1]

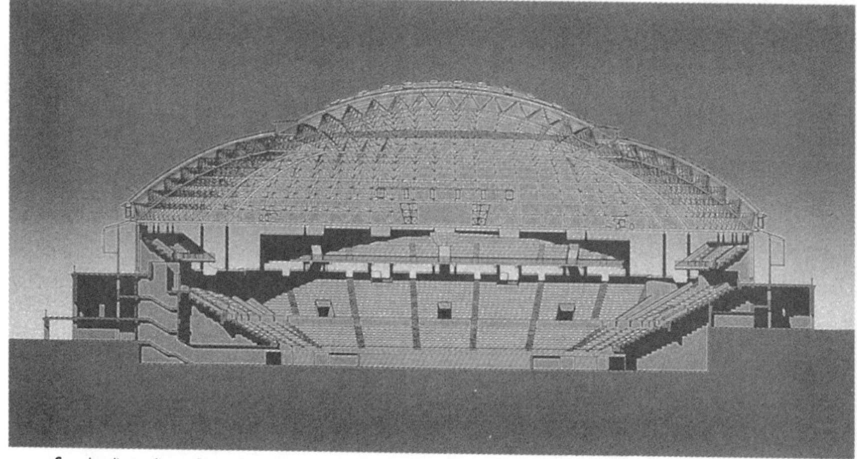

*San Jordi stadium, Barcelona Sports Hall. Photograph courtesy of Arata Isozaki & Associates.*

*How does he get his buildings to speak and what do they say?* Isozaki designs his projects with the individual in mind. He tries to imagine and experience each, not as the architect, but as an individual asking what does the design convey, what does it say? At the essence of each of his designs is a theme or an image. With a theme in mind, his design begins to take on a personality of its own. At the core of each is his untethered imagination— not bound by traditions or trends, yet able to draw freely on a host of shapes, materials and methods from the past and the present. His imagination and the theme of his design are guided by the site's environs —the surrounding buildings, landscape, the life force of the area. By looking at some of his works we gain some insight into his creative process.

• Consider the Sant Jordi stadium he designed for the main event area of the 1992 Olympics in Barcelona. Because the landscape here consists of

---

[1] *Arata Isozaki's Tokyo*, by Arata Isozaki, *Vis à Vis*, September 1990

rolling hills, Isozaki initially envisioned a wavy roof to simulate the landscape. But mathematics and physics showed that a large wavy expansive dome would not be practical. His final design reflects both the gentle slopes of the surrounding hills and a creative engineering feat by engineer Mamoru Kawaguchi. Instead of designing and building a dome in a traditional way, it was first assembled on the floor of the stadium in a folded form consisting of three hinged steel sections, spanning 400 by 350 feet.

*MOCA. Photograph by Yasuhiro Ishimoto.*
*Courtesy of Arata Isozaki & Associates.*

Over a period of 10 days, hydraulic jacks situated on the stadium floor ever so slowly raised and unfolded the dome, like a budding flower, to its height of 148 feet! At this point tension bands were put into place that secured the dome. An incredible sight to observe.

• For the Museum of Contemporary Art (MOCA) in Los Angeles, a small village is the theme. The sunken courtyard is the village square or gathering place of the galleries of this "art village". A variety of geometric forms — curves, pyramid skylights, cubes, a semicylindrical dome make this village stand out like a piece of sculpture among the surrounding buildings in the California Plaza. "Not since the French architectural visionaries of the 18th century has an architect used solid geometric volumes with such clarity and purity, and never with his sense of playfulness," said critic Joseph Giovannini when MOCA was completed in 1986. In addition, in the land of movie stars, it is not surprising that Isozaki was inspired to create his Marilyn Monroe french curve which appears in many of the buildings at MOCA. He often punctuates his works with his humor, playfulness or other personal hallmarks.[2]

*Art Tower Mito. Photograph by Yasuhiro Ishimoto. Courtesy of Arata Isozaki &Associates.*

• Gravity is the theme of the Art Tower Mito complex. Although this redevelopment project is small when compared to the size of Mito, its very design makes it a magnet cultural center, a center for the arts. The buildings are a diverse combination of geometric solids, and the design

---

[2] For example, Isozaki's wonderful sense of humor is evident in his design of the Fujimi Country Clubhouse. Designed in the shape of a question mark, it asks "Why do the Japanese love golf so much?"

of their arrangement and exterior stone facades make an exquisite sight. Since the arts can elevate our spirits, one is struck by two objects at this complex which appear to defy gravity thereby symbolizing the freedom of art. First, the "floating" boulder, in the courtyard area of the art gallery, immediately captivates the observer by appearing to be suspended in air and showered with water. Second, the symbolic tower literally takes one's breath away —a tower of tetrahedrons reaching 330 feet up

*Team Disney Building. Photograph by Yasuhiro Ishimoto. Courtesy of Arata Isozaki & Associates.*

—a helix of tetrahedrons joined in such a way that their tittering appearance seem to defy gravity.

•For the Team Disney Building, time is the underlying image. The courtyard is a gigantic sundial made on a huge truncated cone, with a 120 foot diameter base and 120 feet high. The dial on top of the cone casts a shadow in the cone's interior indicating both the time and the seasons.

Carrying out the time theme further, stepping stones are inscribed with quotations dealing with time from a diverse range of personalities including Einstein, Shakespeare, and Donald Duck. The way the buildings are put together gives off a sense of movement — the truncated cone appears to be intersected by a rectangular structure, a cube skylight intersects the roof at a rackest angle — each suggesting movement, perhaps the movement of time.

•The Joint-Core System is a very impressive visionary plan for an urban forest of residential housing. The shapes of the residences minimize the use of land, yet maximize the use of air space. These unique geometric shapes enhance the sunlight for the living units, and are reminiscent of what could be described as fractal architecture.

These five projects of Isozaki give only a glimpse of his magnificent and eclectic works. Arata Isozaki is one of today's most innovative and creative architects. Viewing the scope of the structures he has designed one can't help but note that there is no challenge which does not capture his imagination. His career of over 35 years is making a lasting impression on the evolution of contemporary architecture.

...number is merely the product of our mind.

—Karl Friedrich Gauss

# fuzzy numbers

The term *fuzzy numbers* sounds like an oxymoron. Aren't numbers supposed to be precise? Isn't 5 of something exactly and always five? In the past, precision was less important. Numbers beyond 2 were often expressed by such words as "few", "many", "several". Ever since people first began using the counting numbers, numbers have been on the go. It seems one set of numbers leads to another, from counting numbers to whole numbers to integers to rationals to irrationals to reals to complex. Add to these all the special subsets, such as even, odd, negatives, amicable, figurative, transcendentals, etc.. How in the

world can the word fuzzy be associated with numbers? Are fuzzy numbers numbers that are vague? Numbers that fall in gray areas? Fuzzy numbers are connected to fuzzy sets,[1] which are linked to fuzzy logic, which may revolutionize how computers work and operate to solve problems.

A fuzzy set *or* fuzzy subset is a group of objects. *Fuzzy numbers* are fuzzy subsets of the numbers on the real number line. The catch is that what elements belong in these sets is not exactly clear. Whether elements even belong to a fuzzy set is not cut and dry. For example, when

---

[1] The origins of fuzzy logic began in 1920 with logician Jan Lakasiewicz , who revised traditional yes-no logic to "multivalent" (multivalued) logic. Then in 1965, mathematicians Lotfi Zadeh applied multivalued logic to set theory and developed *fuzzy sets.* From then on the term *fuzzy* has been applied to more and more things.

referring to the set {all blondes}, does this include bleached blondes? Only blondes at birth? what shades of blonde? A blondish/brown haired person? In other words, there are gray areas. On the other hand, the set of whole numbers {0,1,2,3,4...} is not a fuzzy set. We know exactly which elements belong and which do not. Who is in the set of old people? It depends upon whom you are asking. Ask a 5 year old child, she may say anyone older than 15 years. Ask a centenarian, he may say anyone over 90. The possible number of different fuzzy sets of old people is enormous. Similar cases hold for fuzzy numbers. Is there a fuzzy 1? ... Consider the value of $1 for the IRS. It's $1 whether it is exactly $1, $1.01, $1.003, $0.97 etc.. It's all $1 to the IRS. But 1 and 1.01 are definitely different quantities for a chemist performing certain measurements.

Let's take a closer look at 1 versus fuzzy 1. 1 is 1. It has a specific location on the number line. A segment demarcates its exact location. No other number has that location. That location is *one-hundred percent* 1. But there are numbers, for all intensive purposes, that are 1 but are not. 999/1000 in some cases is as good as 1 itself. The IRS considers 999/1000 dollars, $1. *1 is 1 for everybody, but fuzzy 1s are different for different people.* Take an interval of 1/4 around 1 and call these numbers a fuzzy set of a fuzzy 1. Someone else may take the interval 1/8 around 1 to be their fuzzy set for their fuzzy 1. And so fuzzy numbers are born. Their fuzziness is by *degree* and *is relative*— but for sure fuzzy 1s are never 100% 1. Mathematicians have drawn these fuzzy numbers in different ways on the number line. One way is to use triangular fuzzy

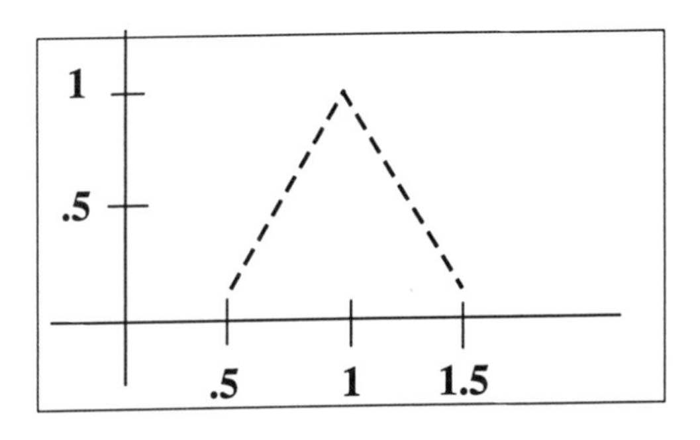

numbers, whose base vertices are the boundary points around an interval of 1 that was chosen arbitrarily. For example, this triangular fuzzy number can be used to represent fuzzy 1.

Among other shapes that have been used to represent fuzzy numbers on the number line are trapezoids and exponential curves. If the fuzzy triangle for fuzzy 1 is skinny enough, it becomes 1. In this fashion, we can discover how regular numbers are a special case of fuzzy numbers. We can subtract fuzzy triangular numbers as we do regular numbers, yet every fuzzy number can be drawn in an infinite number of ways, i.e. there are infinitely many different fuzzy number renditions for fuzzy 1.

What do we do with fuzzy numbers? Mathematicians are developing a whole system of fuzziness. Here we find fuzzy logic[2] as a superset of traditional logic that is designed to deal with partial truths, gray areas. Mathematicians have invented fuzzy calculus, fuzzy differential equations, fuzzy models,..., and yes, fuzzy numbers.

Today's industries, here and abroad, are using and exploring fuzzy systems in the design of "smart" machines, computers, cars, subways, air conditioners, etc. In fact, NASA is planning to use small robotic spheres[3], the size of softballs, to help with various tasks aboard the International Space Station 2002. Where is all this fuzziness going to lead? Time will tell. It's only the beginning.

---

[2] Traditional logic is true or false/ yes and no. Fuzzy logic has a lot of gray areas. For example, if a log is put into a wood burning stove, at what point of its burning is it no longer a log — this is relative. We can't all agree on the moment it is no longer a log. Fuzzy logic is the logic of uncertainty.

[3] Howie Choset of Carnegie Mellon University has been intrumental in developing these robots with NASA.

...they (mobiles) are nevertheless at once lyrical inventions, technical combinations of an almost mathematical quality, and sensitive symbols of Nature, of that profligate Nature which squanders pollen while unloosing a flight of a thousand butterflies, of that inscrutable Nature which refuses to reveal to us whether it is a blind succession of causes and effects, or the timid, hesitant, groping development of an idea.

—Jean Paul Sartre

# *mathematics & the art of* Alexander Calder

In the early 1930s a new art form evolved —*mobiles.* Its creator was Alexander Calder. Alexander Calder (1898-1976) was born into a family of artists. His father and grandfather were sculptors, his mother an accomplished painter. However, he chose to study engineering; a field which influenced the art form he would evolve.

Today, it is difficult to believe that mobiles have not always been around. Or perhaps they have, if we consider the way a branch dangles in the breeze, how leaves hang from a limb, the sway of a palm frond. As he himself points

*A Calder stabile. Esplanade de la Défense. Paris, France.*

out, Calder drew on nature for his models, stating "I felt there was no better model for me to choose than the Universe ... spheres of different sizes, densities, color, and volumes, floating in space, travelling clouds, sprays of water, currents of air, viscosities, and odors — of the greatest variety and disparity."[1] Having been introduced to astronomy as a young boy by his mother, he carried the universe as his theme throughout his work, declaring, "I work from a large live model."[2]. Calder's mobiles speak of nature's dynamic systems, be they the clockwork of the planets and stars, the changes of colors, the movement from the smallest molecules to the ocean's waves. As Jean Paul Sartre describes them,"Mobiles have no meaning, make you think of nothing but themselves. They are, that is all; they are absolutes. There is more of the unpredictable about them than in any other human creation. No human brain, not even their

[1] From the Calder exhibit, San Jose Museum of Modem Art, Nov. 15 – Feb. 1998.

[2] From *Calder* by Ugo Mulas and Howard Arnasar, page 45. The Viking Press, New York, 1971

*A Calder mobile. East Building of the National Gallery of Art, Washington, D.C..*

creator's, could possibly foresee all the complex combinations of which they are capable. A general destiny of movement is sketched for them, and then they are left to work it out for themselves. What they may do at a given moment will be determined by the time of day, the sun, the temperature or the wind..."[3]. Although the mathematics of complexity theory at this time had no formal labels, these mobiles illustrate the themes of the balancing act between chaos and order. Sartre describes such a tug-of-war between chaos and order: "Suddenly a mobile beside me, which until then had been quiet, became violently agitated. I stepped quickly back; thinking to be out of its reach. But then, when the agitation had ceased and it appeared to have relapsed into quiescence, ... These hesitations, resumptions, gropings, clumsinesses, the sudden decisions and above all that swan-like grace make of certain mobiles very strange creatures indeed, something midway between matter and life."[4] It is no wonder that Calder's

*A Calder mobile-stabile, displayed in the Senate Building, Washington, D.C..*

---

[3] From *Calder* by Ugo Mulas and Howard Amasar, page 45. The Viking Press, New York, 1971

[4] *Ibid.*, page 47.

mobiles display forms of complexities. The design of a mobile has inherent in it many physical forces at work and at odds — multiple centers of gravity (found in each of its components and that for the entire piece), — centri- fugal forces caused by swaying — periodic cycles of movements — speeds and accelerations — vectoral quantities — centers of differing densities — all of which Calder had experienced in problems of engineering.

Although Calder worked in other media, he is renowned for his sculptures, be they *mobiles* or *stabiles*.[5] His mobiles seem to float effortlessly in the air. They seem so much a part of the space they occupy that one does not consciously search for their connection, but instead is taken in by its subtle movements as affected by the slightest breeze or vibration. It is all reminiscent of the connection between Foucalt's pendulum and the Universe. The shapes and movements traced by the mobiles are patterns we have experienced inherently. Even his *stabiles* — his massive stationary metal sculptures — project movement through the shapes of their arcs and curves. The sheer magnitude of these curves move as we move around them and view them from different perspectives. We in essence become a mobile part of the stabile.

Calder's work touches everyone. Albert Einstein gave the ultimate compliment. After being captivated by the Calder mobile, *A Universe,* he declared he regretted not having thought of it himself.

---

[5] In 1932 Marcel du Champ named Calder's work *mobiles* and the same year Jean Arp named Calder's *stabiles*.

Nothing in nature is random...A thing appears random only through the incompleteness of our knowledge.

— Spinoza

# wavelets —

## *a mathematical way of describing sound and other things*

During the first four months of gestation, fingerprints, one's identity markers develop. They even differentiate you from an identical twin. What do fingerprints have to do with mathematics?

The mathematics used to describe fingerprints originated with the mathematical study of sound. 19th century French mathematician Jean-Baptiste Fourier ingeniously devised sinus curves (infinite periodic functions) to describe and replicate sound. Regardless of the complexity of the sound, a finite number of different sinus curves could be combined to describe any particular sound — be it a musical note or the sound of a falling tree. Sine waves rely on the amplitude and frequency of the sound to create a graph and equation of the sound. When a note is played it vibrates at a specific frequency and has a particular amplitude. These two characters — frequency and amplitude — remain with the note and thereby with the curve until it dissipates. This explains why sinus curves continually repeat the same wave shape for a particular sound. Over the years these sinus curves have been used to describe other things besides sound, and in doing so mathematicians have adapted and created an area of mathematics referred to as Fourier analysis. Fourier analysis uses sinus curves and sinus equations to describe radio signals and heartbeats, to design the optimum structure and fabric for a particular instrument. These waves can be used to characterize a single sound, such as the note A or a conglomeration of sounds. They do so by combining the waves for the individual sounds into a new sinus curve (see graph). This method has its drawbacks because the information of each individual sound wave ( i.e. the amplitude and frequency of each) is lost in the final curve. For example, suppose you have the problem 3+5=8. The final result is 8. If I just gave you 8 and asked what numbers did I add to get 8, the possibilities are infinite—1+7; .3+.7+5+2 etc. So in a room full of noise created by a variety of means, the sinus waves describing the noise do not identity the individual sounds composing the noise or the particular sounds that occur. To improve upon the information possible from the Fourier waves, mathematicians decided to introduce time into the picture by breaking down the wave into smaller fixed intervals. Thus each interval, called a *window*, has its own sinus

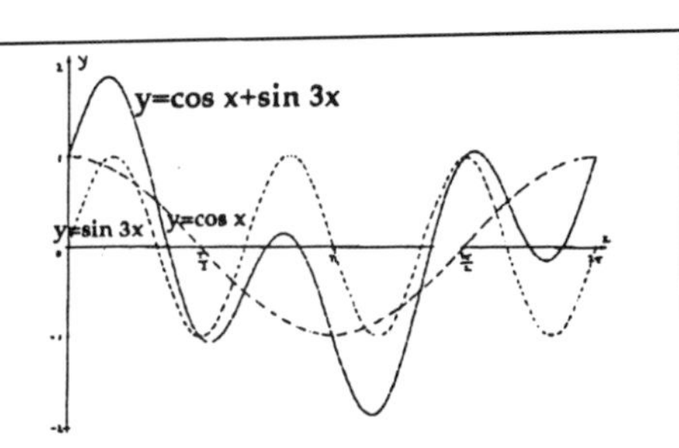


*This graph shows two sinus curves, y=cosx and y=sinx, and their result when combined to form the equation y=cosx+sinx.*

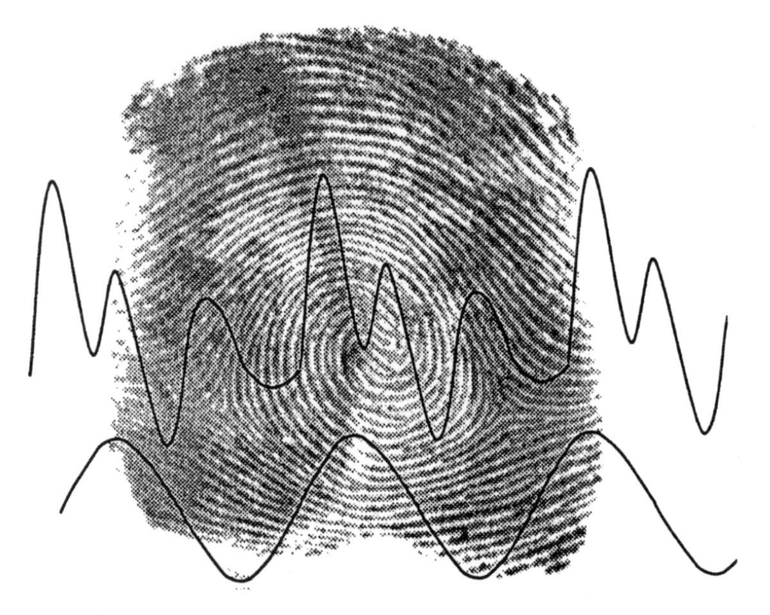

wave describing it. When a window has only its sound, characteristics are preserved within that window. But, within a window a number of different sounds often occur, and a single sinus wave is still used for the entire window. In other words, the individuality of sounds is still not preserved in a window of time. In addition, a particular sound may overlap windows, and again its individual wave would have been lost. These windows are all of fixed length, analogous to the measures of a musical score. Yet these mini waves of windows are again insufficient to describe every sound component of a room.

A major breakthrough came in 1987 with the wavelets devised by Ingrid Daubechies. Daubechies introduced time in a new way into the graph of the waves. She connected the pitch of the sound to time. The higher the pitch the shorter the wavelet. Thus, each wavelet describing each sound is a single curve unto itself, yet a string of wavelets can be used to describe a conglomeration of sounds or objects. These wavelets are like fractals of sound— the further away you stand from the wave the general shape of the sound of the room appears, but as you zero in on the wave the individual wavelets come into view. The Fourier mini waves are periodic, they repeat their shapes. Yet, Daubechies' wavelets are not repetitious.

How are these wavelets connected to fingerprints, heartbeats, radio signals, static noise, earthquakes? Anything that can be described or appears as a picture or sound (any vibration) lends itself to be characterized by wavelets. Ever wonder how ink thumbprints have been replaced by a screen print at the DMV? Now one's fingerprint is broken down into a series of wavelets with particular amplitudes, frequencies and lengths. The fingerprint's wavelets are translated into a numerical description (equations and values) which can be stored and transmitted electronically via the computer. Initially a standard fingerprint required a substantial amount of computer data space.[1] But an ingenious way of compressing the information has been developed which takes up 1/20 the computer space. This process dramatically facilitates identifying and matching prints.

Hand in hand with the wavelets and voiceprints is the field of *biometrics.* Here mathematics quantifies physical and behavioral characteristics. Besides fingerprints, other traits include the eye's retina prints, voice prints, DNA prints, geometric analysis of the hand.

The ramifications of mathematizing everyday phenomena are far reaching, and mathematical footprints continually surface in such diverse fields as medicine, quantum mechanics, seismology, criminology, computers and cryptography?

---

[1]The fingerprint is represented as a combination of wavelets by looking at it as a 2-dimensional function. Jonathan Bradley and Chris Brislawn at Los Alamos National Laboratory working with Tom Hopper at the FBI devised a wavelet standard for fingerprints which compresses the data from 10 megabytes to only 500 kilobytes. These electronic fingerprints are near perfect matches with the original prints.

# the longitude problem

> . . .Alas, through a mystery of nature, every means conceived to establish longitude has always proved faulty.
>
> —Umberto Eco
> *The Island of the Day Before*

What do Captain Cook, Captain Bligh, Darwin, Newton, Galileo, Halley, Cassini, Huygens, Hooke and Flamsteed have in common? They, and many others, were all in some way connected with the elusive longitude problem, whose solution would have a profound effect on our lives. Without an accurate and practical way to determine longitude, explorations of new worlds would have remained haphazard, dangerous and futile.

Where am I? Where was I? Where am I going? may sound like philosophical questions, but these are questions vital to voyagers.

The answers depend on an unusual type of coordinate system whose origin dates back to c. 276 B.C. to Eratosthenes, who placed the first lines of longitude on a map of the Earth. Over the centuries lines of latitude and longitude checkered maps of the then known world. For example, around 150 AD Ptolemy made his world atlas, called *Geographia.* His maps included both latitude and longitude lines. Yet the mere knowledge of a location's coordinates on a map did not mean seamen knew how to get there or even how to get back. In the time of Galileo there were no Global Positioning Systems with sophisticated satellites and radios to almost instantaneously determine and relay information of a boat's position. Sailors had to rely on the stars, crude instruments, good weather and intuition, all of which were iffy. Seamen, merchants, explorers, kings, scientists knew the economic and political importance of being able to find one's location at sea. Many horror stories of ships lost at sea, lives given up to starvation, scurvy, or drowning became legends. For example, in a 1741 voyage Commodore George Anson lost his way, and sailed back and forth in search of Juan Fernandez Island. His errors resulted in the loss of many crew members to scruvy. Even the most experienced captains wandered off course for long periods of time in the vast expanse of ocean in search of land, such as Admiral Sir Clowdisley Shovell who, in 1701, made a disastrous misjudgement in longitude which caused the loss of four of his five ships. Others encountered the ambush of pirates waiting along the trafficked routes.

Why couldn't a global coordinate system of latitude and longitude lines guide these seafarers? Using the stars and instruments of elevation sailors knew how to determine their latitude[1] . The problem was with determining the longitude. The lines of latitude are the parallel circles drawn above and below the equator. The equator's latitude is arbitrarily designated as 0°, the north and south poles latitudes are then 90°N and

---

[1] Latitude is determined with respect to the equator (any point on the equator has 0° latitude). The North Pole's latitude is 90° N of the equator and the South pole is 90° S of the equator. You can estimate latitude by pointing the index finger of one arm at the North Star and the other at the horizon, and measuring the degree of this angle. This is the degree of your latitude.

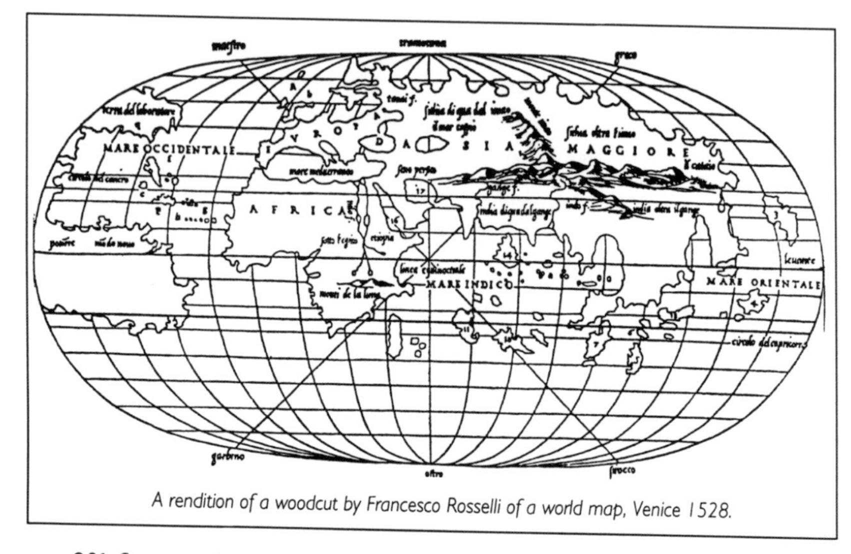


*A rendition of a woodcut by Francesco Rosselli of a world map, Venice 1528.*

90° S respectively. Longitudes, on the other hand, are related to the great circles of the Earth which pass through both of its poles. They are oriented to a 0° longitude line (actually a semicircle) called the prime meridian, whose location is purely arbitrary. Longitude is a line (semicircle) so many degrees east or west of the prime meridian. These semicircular lines are also called meridians[2], and today the prime meridian passes through Greenwich, England. Over the years the prime meridian has had various locations. In 1884, 26 countries met to fix the prime meridian at its present day location at Greenwich, England and to establish a time zone system. Other locations that have been prime meridians over the years include the Azores, Cape Verde Islands, Rome, Copenhagen, Jerusalem, St. Petersburg, Pisa, Paris, and Philadelphia. Imagine the Earth covered by meridians, 24 of which divide the globe into its 24 time zones. 24 meridians (longitudes) section the globe establishing 24 times zones of 1 hour each. Along the equator the distance between two adjacent meridians is the widest. As one moves north or south

---

[2]Two meridians forming a great circle are called a meridian circle. *Meridian* comes form the Latin *meridies*, meaning middle of the day or noon. Wherever you are, regardless of the season, when the sun is directly over your head, it is noon everywhere along that meridian passing through your feet from North to South pole.

along a meridian the distance between two adjacent meridians gradually diminishes until it is 0 at each pole.

Seafaring countries engaging in commerce were well aware of how crucial the solution of the longitude problem was. In fact, over the years France, England and Spain had offered prizes as incentives to solve the longitude problem. A myriad of proposed solutions had evolved over years. Among the many methods proposed to measure longitude were—

1) The magnetic variation method used the compass to ascertain the difference between the reading of magnetic north and the North Star for true north and splitting the difference of these two to get longitude. But variations in readings fluctuated due to one's location making this method inaccurate. This phenomenon, variations in the magnetic compass as longitude changes, was first noticed by Christopher Columbus on his first voyage to the Americas in 1492.
2) Dead reckoning method.
3) Powder of sympathy method was a bogus method that relied on a powder that was supposed to heal wounds over distances. It was supposed to be used to establish time differences between the home port and the ship's time by applying powder on an old bandage of a wounded dog, which was now on the ship. The powder, applied at noon on the bandage which was left at the home port, would instantaneously affect the dog on the ship and tell the ship's captain that it was 12 noon at home port.
4) Moons of Jupiter method presented by Galileo to the Spanish King Philip III, who had offered in 1598 the prize of a lifetime pension for the answer to the longitude problem. Galileo's method required the sailor to observe Jupiter's moons and locate them on tables Galileo made. The difficulty in viewing these moons especially aboard a ship, did not make this astronomical solution feasible. In the mid 1600s work on this method was renewed in France by the French Académie Royal des Sciences.

5) 1514 German astronomer Johannes Werner felt that longitude could be determined by charting the position of stars and the Sun along the path of the Moon. In theory it seemed plausible but astronomers at this time were unable to predict the course of the moon on a daily basis nor were there accurate instruments that sailors could use to measure the distances between certain stars and the moon. Around 1674, King Charles II of England asked his experts (Robert Hooke, John Flamsteed, Christopher Wren) to evaluate the resurgence of this method by Frenchman the Sieur de St. Pierre. This resulted in the establishment of the Observatory at Greenwich, where Flamsteed spent 40 years making celestial charts for navigations. His catalog was published after his death in 1725.

But none had actually solved the problem. Scientists such as Galileo, Newton, St. Pierre, Cassini, and Flamsteed felt the solution lay with astronomy. Some people felt the answer was more easily solved by developing an accurate and reliable timepiece. In fact, in 1533 geographer Reiner Gemma Frises explained how longitude could be calculated using a mechanical clock. But at this time there were no clocks that could keep time to the accuracy needed for this method to work. Others came up with some rather bizarre proposals.

Ongoing shipping disasters eventually led to outcries from merchants, sailors and scientists, and prompted England to enact the Longitude Act of July 8, 1714. Sailors needed more than star maps, crude tools or dead reckoning to find their way. They needed a consistently accurate means that would work under any conditions. The act established a Board of Longitude and a £20,000 first prize (≈$1,000,000 in today's standards) to the first person to devise a means for determining longitude to within an accuracy of 1/2 degree.[3] Who would solve the longitude problem

[3] Longitude, time and distance are all related on the globe. The accuracy of 1/2° means an accuracy of 32 land miles along the equator which diminishes proportionally as one moves north/south along a longitude line toward either pole.

and claim the prize? Unfortunately politics, intrigue, scandals[4] and egos entered the picture. Many scientists, among them Isaac Newton, John Flamsteed, and Nevil Maskelyne, felt a mechanical device, such as a clock, would not be a reliable solution, and the only possible method lay with celestial calculations and mappings. To this end, John Flamsteed, the first astronomer royal at Greenwich, spent 40 years charting the night skies and compiling almanacs of tables and charts.

John Harrison, a self taught clockmaker, decided to tackle the longitude problem. He felt if he could design a timepiece that would be impervious to the elements of an ocean voyage and could accurately maintain the time of the port of departure, then longitude could easily be calculated in the following way.

*The clock on board a ship would maintain the time of the port of departure (or the time of the prime meridian or any other place of known longitude. and time). To determine the ship's longitude anywhere along its course, one simply needed the difference between port time and the ship's time. For example, the time difference for 32 minutes and 20 seconds would be calculated as follows:*

- *The globe is divided into 24 time zones of one hour each, and also into 360°.*
- *So 24 hours = 360° of longitude*
- *1 hour = 15° of longitude*

---

4 • Edmund Halley was tired of waiting for John Flamsteed, the astronomer royal, to give his okay to publish his star maps. Consequently, Halley took it upon himself to secretly print Flamsteed's maps without permission. Flamsteed was furious and retrieved and destroyed as many of the 400 copies as he could find. Flamsteed did not feel his charts were ready for publication.

- Robert Hooke and Christian Huygens argued over who was entitled to the English patent for a clock design to which they both claimed ownership.
- Nevil Masklyne, an astronomer royal, repeatedly tried to thwart John Harrison from claiming England's longitude prize. Instead, he tried to promote his own inaccurate astronomical solution.

- *Hours of time are divided into minutes and seconds of time.*
- *Degrees of longitude are divided into minutes (symbol is ') and seconds (symbol is ") of longitude or distance.*
- *Thus 60 minutes of time = 15°x60=900' of longitude.*
- *So 1 minute of time = 15' of longitude. Carrying this further, we get*
- *60 seconds of time = 15'x60=900" of longitude, so 1 second =15" of longitude.*
- *Thus, in this example, the difference of times between ship's clock and the home port, namely 32 minutes 20 seconds converts in the following way to a longitude reading—*
- *(32 minutes)(15')= 480' which converts to 8° of longitude by 480'/60*
- *(20 seconds)(15") =300" which converts to 5' of longitude by 300"/60*
- *So this ship's longitude at this point is 8°5' east of the port because it gained time. If the time was 32 minutes 20 seconds less than the port's time, it would have been 8°5' west of the port.*
- *Combining the ship's longitude with its latitude reading pinpoints it location on the map.*

Harrison determined which materials would be self-lubricating, what to do to compensate for temperature changes, and how to wind a clock without losing time. His clocks were works of intricacy and precision. He was the first to design a clock that could survive ocean voyages and comply with the accuracy demanded by the Longitude Act. Started in 1727, his work on the longitude problem and his clock designs consumed 46 years of his life. For Harrison, his work had been a labor of love, a labor of obsession, a labor of perfection. Over the years he presented five maritime clocks[5] to the Longitude Board. His H-4 (his fourth timepiece) went on its sea trials in 1761, and it performed exquisitely — being accurate to within 1/50 of a degree. Did the Longitude Board award him his due prize? No! Instead, they changed rules and added

[5] His first clock weighed about 72 pounds, while the last one he designed was the size of a pocket watch.

conditions that they claimed were not met by H-4. They gave Harrison £ 1500 as a conciliation, and insisted H-4 would have to pass yet a second trial. Over the years the Board had given him small stipends to supplement his work, but when he deserved the prize, they pulled back. Now an old man, Harrison was determined to seek a resolution. In 1773, after making his fifth timepiece, he enlisted the support of King George III, who was sympathetic after hearing his story and testing his superb clocks. The king went directly to the Parliament, which granted Harrison £8750. Unfortunately, it was not the Longitude Board that made the award. In 1773 the Board adopted a new Longitude Act with convoluted conditions, and the prize money was never awarded. Harrison's work and innovations in clock making were essential to the clock making revolution that followed after his death in 1776.

Like many mathematical problems, the search for a longitude solution spawned other important discoveries along the way. Pendulum and water clocks were replaced with balance spring clocks, Galileo discovered the clockwork in the moons of Jupiter with his ephemerides, Danish astronomer Ole Römer discovered in 1678 that light did indeed have a measurable speed. Römer used his celestial observations and calculations to calculate the speed of light as 240,000km/sec. Today, its actual speed has been set at 299,792.458 km/sec.

Problems such as the longitude problem have acted as catalysts to advance our technologies. Today, we have come a long way from sand clocks to atomic clocks and our sophisticated tracking systems and radar have replaced the old solutions to the longitude problem. Mathematics was an important player in its solution.

# mathematics in the creases

It must have required many ages to realize that a brace of pheasants and a couple of days were both instances of the number two.

—Bertrand Russell

Ever think that a creased sheet of paper would conceal mathematical ideas? Just look at this folded square.

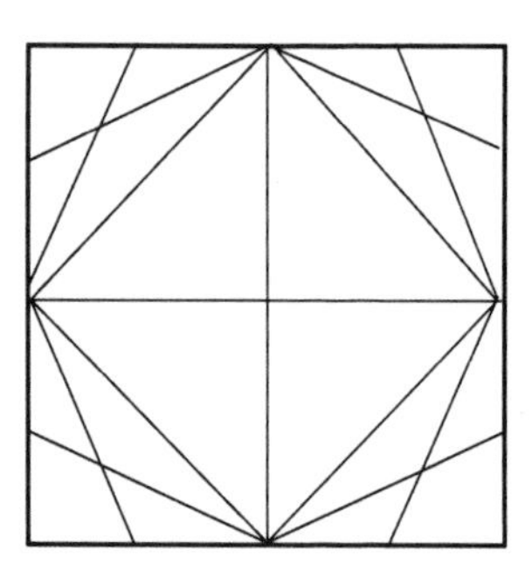

What do you see?
Do you see an inscribed square?
Can you make out a regular octagon?
What about this figure?

Can you see how its creases demonstrate that the three angles of a triangle total 180°?

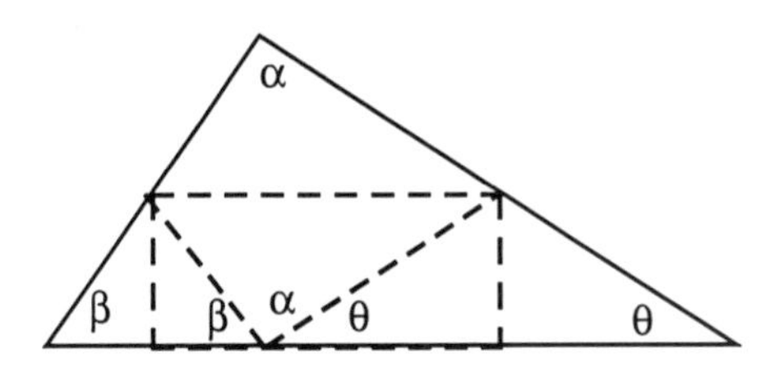


Now consider the swan origami figure on the following page.
It is unfolded into its essence —the flat square piece of paper.

Study the creases. What do you see?

- Note the symmetry of the lines.
- Note the repetition of a pattern.
- A closer inspection reveals a self-replicating pattern, reminiscent to the mathematical eye of fractals.

By folding the square and creating an origami figure you can transform a 2-D object into a 3-D object — in a sense a square piece of paper is made elastic, which is what is done in the field of topology[1]. Just think — any point on a piece of paper has the potential of a 360° angle.

These examples illustrate some of the hidden mathematics in paperfolding. Mathematicians are notorious for playing with objects that the layperson considers recreational or a pastime— soap bubbles, origami, flexagons, paper models of poly-hedra, magic squares, and tessellations to name a few. But to the mathematician these objects provide a means for exploring and discovering mathematical ideas and

---

[1] Topology is a field of mathematics that studies properties of objects that remain unchanged after they have been transformed by various means, such a stretching..

perhaps help in devising new solutions or proofs to problems which may seem totally unconnected to *the pastime at hand.*

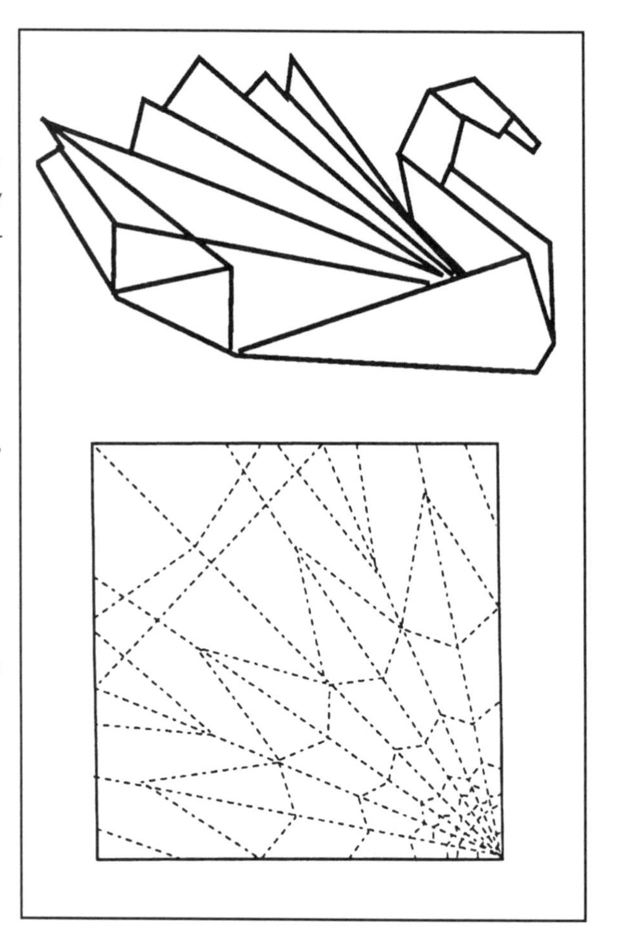

Today, we find mathematicians, scientists and engineers exploring high powered mathematics in paperfolding, as witnessed at the 2nd International Meeting of Origami Science & Scientific Origami (IMOSSO) in November of 1994.

Koryo Miura, professor emeritus at the University of Tokyo, and Thomas C. Hull of the University of Rhode Island were among the many mathematicians at the IMOSSO conference. What have they discovered using paperfolding?

•Miura is the creator of the *Miura Ori* folding pattern which was used in the retractable solar panels mounted on Japan's Space Flyer Unit artificial satellite. His special way of folding a rectangle has also been used in map folding. A map folded in the *Miura Ori* manner can be easily opened and closed as a whole or in sections.

•Hull studies the geometric connection between origami forms and their crease patterns. Using mathematics found in graph theory together with his work on the creases of origami patterns, he has expanded on ideas of other mathematicians and proven numerous theorems. His work has enhanced both origami creations and his research in planar graphs. In addition, he explores and studies the crease patterns of origami as well as tessellation patterns by looking for both periodic and nonperiodic tiling patterns. As he says, "There are lots of interesting problems here, with links to other kinds of mathematics."[2]

•Mathematician William T. Webber is another paperfolding mathematician. He has used the mathematics of tessellations and paperfolding to transform a flat rectangular sheet of paper into a

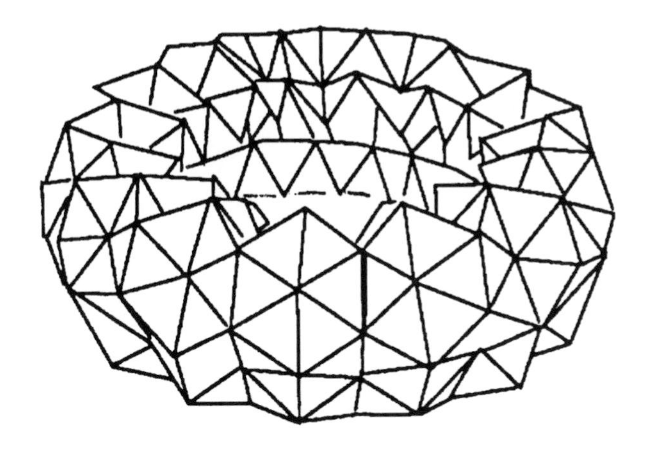

[2]*Paperfolds, Creases, and Theorems* by Ivars Peterson. *Science News*, January 21, 1995.

> phenomenal 3-dimensional solid, called a toroidal polyhedron (a 3-D ring whose multiple faces are triangles). Was Webber satisfied with one such object? No! He has experimented with different shaped triangles and folding patterns. In fact, he has discovered infinitely many families of these paper rings with distinct properties. Thus far he creates these toroidal polyhedra by folding along edges of isosceles triangles drawn in a tessellating pattern on paper. He uses the computer to determine whether a particular model is indeed a toroidal polyhedron by calculating the angles where the triangles meet. As of 1997 he has not found a pattern of equilateral triangles that yields a toroidal polyhedron. Only time will show what treasures these mathematical rings will unfold.

Paperfolding as an art form has been evolving for centuries. Paperfolding as a mathematics form is a 20th century phenomenon. The beauty of the mathematics of paperfolding is how it continually mingles art and science, allowing the creativity of both to emerge and grow in a variety of ways.

*...and nature must obey necessity.*

—Shakespeare
*Julius Caesar,*
*Act IV scene 3*

# mathematics & nature's formations

The strange yet beautiful rock formations of Monument Valley in Arizona, the rock citadels of Meteora in Greece, the erratic coastlines of California, the spiral formation of the Milky Way galaxy — all demonstrate the tug of war between the elements of nature. All are describable by mathematical ideas. One way to look at mathematics is as the study of patterns — patterns from which we make hypotheses, draw conclusions and seek proofs. Recurring patterns in nature offer a wealth of mathematical ideas to explore and connect to those patterns. Among these are:

• patterns in the petals of flowers and the way leaves grow on stems led mathematicians to recognize the presence of the Fibonacci sequence in such flowers as cosmos and trillium, the branch of an elm, the spirals of a pineapple.

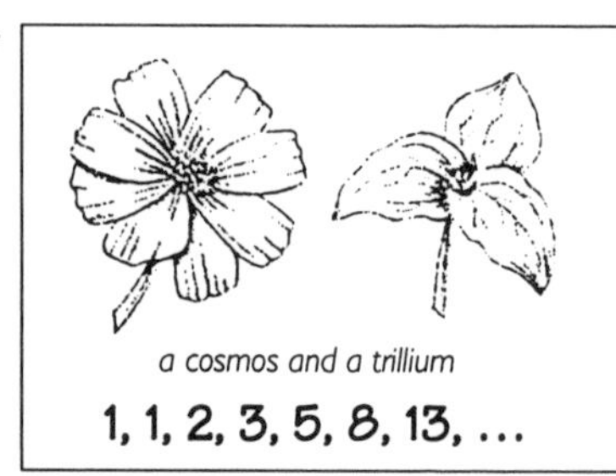

*a cosmos and a trillium*

**1, 1, 2, 3, 5, 8, 13, ...**

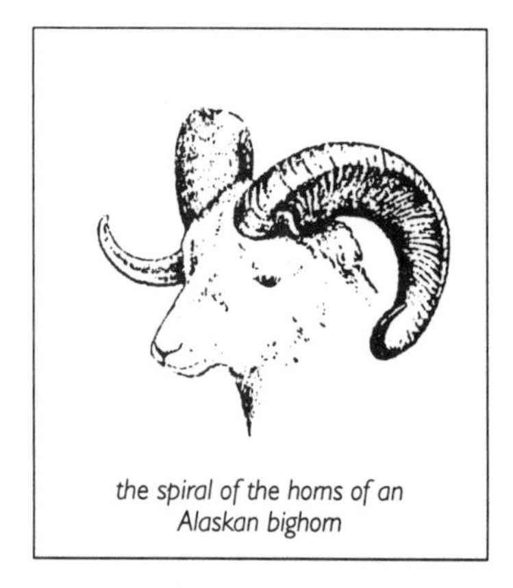

*the spiral of the horns of an Alaskan bighorn*

• recurring curves seen in the antlers of the kudu, the helix of the honeysuckle, the swirl in a seashell, the spiral of a galaxy, the twist in DNA illustrate how spirals can accommodate growth.

• shapes of polyhedra formations found in crystals, rock layers and the facets of gems led scientists to classify crystals with respect to these forms and their symmetries.

*Monument Valley, Arizona*

*Devil's Postpile National Monument, California. Photo courtesy of Margaret Goodrich.*

- studies of the twists, bends, turns of rivers, creeks, and streams seemed to defy logic — their shapes showing that the shortest distance between two points of a stream is not necessarily the one streams take and revealing that their natural inclinations are governed by centrifugal forces, uniform expenditure of energies, and can be described by randomness and probability.
- formations seen in the cluster of soap bubbles, the scales of an armadillo, the cracks in a mud floor, the kernels on a cob of corn, the bark of certain trees, the pattern of a giraffe's hide, the cells of honeycomb, the shapes of rocks at Devil's Postpile National Monument, California — reveal evidence of hexagons, triple junction, and ideas of close packing.
- patterns on a wind blown sandy beach, in the shapes of clouds, in the forms of lava flow, of the terrains of mountain ranges, in the

repetition of a design in a fern's leaves — all can be recreated by a mathematician using a computer and fractal equations. Today the theory of fractal mathematics has evolved to new levels via the use of the computer, and new applications and connections to nature are being discovered continually. Fractal forms have come a long way from their early explorations in the 1800s.

• the structure of the fibers of silk fabric resemble triangular prisms which gives silk its shiny look.

What can one do with mathematical descriptions of nature? Since we can see nature with our eyes, take photographs, why do we need equations to describe its forms? Two important areas of mathematics are probability and statistics, which are often closely tied to the field of mathematical modeling. Equations not only describe a shape, but can predict different shapes when input changes. The evolution of a geological region can be predicted as certain changes are hypothesized. What could have been different about ... a flood ... a drought ... an earthquake ... a volcanic eruption ...a dam ...a fire? These hypotheses teamed up with mathematics, computers, and mathematical modeling can be used to predict a range of outcomes. It is here where the study of small changes (often seemingly imperceptible changes in the initial input) may produce drastic outcomes, as proposed by chaos theory and complexity. Words speak of the beauty and wonder of nature's formations. Mathematics gives us windows into its evolving future and how our actions may impact the outcomes.

*Polyhedra formations in a cluster of crystals*

# *mathematics & the architecture of* **pyramids**

Through space the universe grasps me and swallows me up like a speck; through thought I'd grasp it.

—Blaise Pascal

The grandeur of the Egyptian pyramids on the west bank of the Nile River at Giza today greets hordes of tourists, who stare in awe and are mesmerized by their size. How these enormous structures were built without the advantage of modern technology is impressive. Given modern techniques and materials, today's buildings can take on a variety of shapes, layers and forms — witness the hyperbolic paraboloid of St. Mary's Cathedral in San Francisco or the tesseract like structure of the Grand Arche at La Defense in Paris, France. Even so, the pyramid shape still finds its way time and again into the drawings of modern

*A pyramid theme is carried out in the design of this modern office building in Foster City, CA*

architects, adding that recognizable form to the vast architectural landscapes of our cities' skylines. We find pyramids gracing tops of buildings, such as the Transamerica building in San Francisco; acting as windows of light as at the Louvre in Paris; and even as hotels designed to conjure up the Mayan pyramids in Mexico's Yucatan.

What mathematics and utilitarian advantages does the pyramid shape offer? In ancient times, stability of the structure was a deciding factor. The Egyptians did not have access to prepounded concrete and steel to reinforce their structures. They needed a design that was naturally stable. Familiar with triangles, its 3-dimensional extension was natural. But what shaped base was best suited for their means of construction? The Egyptians knew how to make right angles by using knotted ropes and the Pythagorean theorem. They probably relied on the 3-4-5 right triangle and rope stretchers to create the right angles of the square which was the base of their pyramid.[1] In addition, the pyramid shape did not require any interior supports to hold up the structure. And so the square based pyramid was a natural structure for them to select.

Does the pyramid offer any advantages for today's building designs? From a mathematical viewpoint the base of the pyramid can be any polygonal shape. But some shapes adapt more readily to our lifestyle than others. Most furniture works best with right angled walls. Hence, a square or rectangular based pyramid is more practical for living quarters. With architectural skylights and accents, any shaped base is feasible, the deciding factor being taste. With today's various metal compositions, modern pyramids can be framed from lightweight material instead of the 5000 pound stone blocks the Egyptians used. The frames can be manufactured in units and assembled at the site in a semi-prefabricated manner. Today's pyramids do not have to be solid to maintain their stability. Interior supports free the inside of the pyramid to open space, making available light and outdoor living areas connected with interior stairwells, atriums, and different staggered levels which allow natural light to flood and filter through. What is equally wonderful about each level of the pyramid is that outdoor balconies along the lateral faces do not cast shadows on the units below. In addition, pyramid structures deflect wind upward rather than downward. But the pyramid is not without disadvantages. Because of the way it tapers upwards, its inherent shape does not maximize the possible square footage a building with that base and height could have. Euclidean geometry explains how to find the volume[2] of a pyramid— illustrating why a pyramid's volume with the

---

[1]During ancient times there were three possible ways of creating a right angle — making a perpendicular bisector of a line segment using only a straightedge and compass; using knotted ropes to form triangles with sides 3, 4, 5; making a semicircle from a circle and then choosing any point on the arc of the semicircle as a vertex of a triangle whose base is the circle's diameter.

[2]The ancient Greek mathematician Eudoxus (circa 400-337 B.C.) proved the formula — volume of a pyramid = 1/3 base•height. He was from Cnidus and was renowned not only in mathematics and astronomy, but in medicine, philosophy, rhetoric, geography, and law.

same shaped base and height of a prism is exactly one-third that of the prism's volume. On the other hand pyramid structures do not have to end in a point, since they can be truncated and thereby minimize the loss of square footage. Other formulas show that less building material is needed for the lateral faces of a pyramid than for those of a prism.

Today the pyramid is one of the carryovers of the genius of the ancient world. After 4500 years it has not outlived its usefulness, but remains a mysterious and powerful form for some and a familiar shape for others.

*The glass pyramid at the Louvre in Paris, France*

# cellular automata—

*the pixels of life*

Out of nothing I have created a strange new universe.

—János Bolyai

What do dots on a computer monitor have to do with life? Looking at this string of numbers — ...00100000... and applying the simple rule: *after each moment of time a digit is replaced by the sum of itself and the neighbor on its right* — become ...01100000... ; one wonders what's so difficult, so mysterious, so amazing about this? It seems an elementary process, yet in mathematics seemingly simple concepts may end up having complex and far reaching ramifications. This simple example illustrates one step in the life of a *cellular automaton.* The word *cellular* implies some basic or essential unit. *Automaton* indicates a machine or robot. In essence, a cellular automaton is composed of discrete[1] cells which are occupied by discrete values which follow a specified *rule(s)* at each specific interval of time. The discrete cells can be the pixels, the tiny square dots on the screen of a computer monitor. The values, for example, can be the binary digits 1 and 0 which convert to black and white square dots on the computer monitor. The rule can be a computer command (i.e. a program) that specifies how the pixels change from black to white or vice versa. Cellular automata can be 1-dimensional, 2-dimensional, 3-dimensional,..., n-dimensional. A 1-dimensional cellular automaton can be depicted as a string of 1s and 0s *or* a linear chain of black and white pixels which instantaneously change according to a specified rule applied as each interval of time passes. A 2-dimensional cellular automaton can be visualized as a square shaped grid of pixel spaces on a computer monitor. In the 1970s, mathematician John Horton Conway came up with a 2-dimensional cellular automaton called the

*initial cells at 0 sec.*
*after 1 sec.*
*after 2 sec.*
*after 3 sec.*

*a neighborhood*

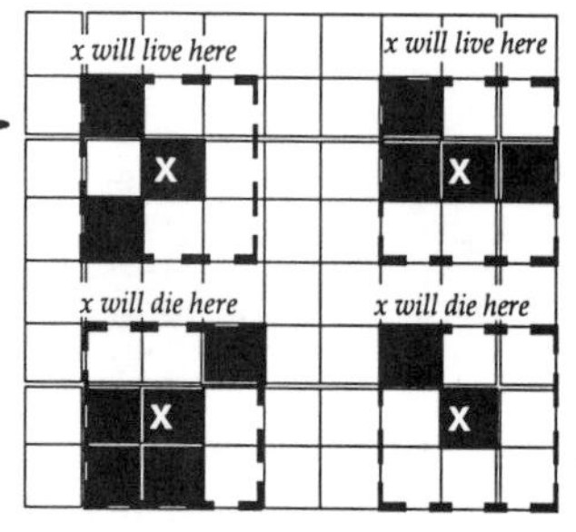


[1] Discrete here means separate/distinct. Two objects are discrete if between any two adjacent ones, no other exists. For example, the natural numbers are discrete because there is no other natural number between two consecutive ones. Pixels are discrete because there is no other pixel between two adjacent ones. It is somewhat analogous to rational numbers versus whole numbers. Between any two rational numbers, another can always be found simply by taking their average. On the other hand, no other whole number exists between any two consecutive whole numbers.

*game of life.* In *the game of life* a cell is designated as dead (0 or white pixel) or alive (1 or black pixel), and the evolution of a cell is dependent on the neighborhood of the cell, which consists of a 3x3 square lattice of cells(pixels) around it. A cell remains alive as long as two or three of its neighbors are alive. It dies if it gets crowded by four or more neighbors or feels deserted when there are less than two neighbors left. On the other hand, a cell comes back to life if exactly three cells become alive in its neighborhood. The *game of life* can produce some fascinating patterns and results as a cluster of initial cells begin their evolutionary process. These are just two examples[2] from infinitely many possible cellular automata. For years the phenomena of universe have been observed and described on a large scale, taking a broad encompassing look at things. In other words, the big picture analysis. The mathematics used to explain the big picture has relied heavily on continuous mathematics, which focuses on concepts from calculus. Now some mathematicians are zooming into the microscopic picture of things, and many of the tools they use rely on those found in discrete mathematics, fractals and nanotechnology. Here is where cellular automata reside.

The idea of cellular automata originated with Konrad Zuse and Stanislaw Ulman in the 1940s. Using the concept of cellular automata John von Neumann showed how a computer could be programmed to self-reproduce[3]. Then in the 1970s, interest in cellular automata was renewed with the premiere of *the game of life*, and interest was boosted further in 1980s by the work of Stephen Wolfram. Today, work in

---

[2]Some other examples of rules that have been used are:

*the majority rule:* here the center cell life depends on the what is the situation of the majority of cells in its neighborhood.

*NWSE-neighbors:* these rules depend on what is the situation with the cells immediately north, west, south and east of the center cell.

*one-out-of-eight-rule:* a cell comes to life if exactly one of its neighbors is alive.

[3]Neumann abstracted from nature the logical sequence of the reproduction process using 2-dimensional cellular automata with 29 possible states per cell, and explained how an abstract pattern could reproduce itself.

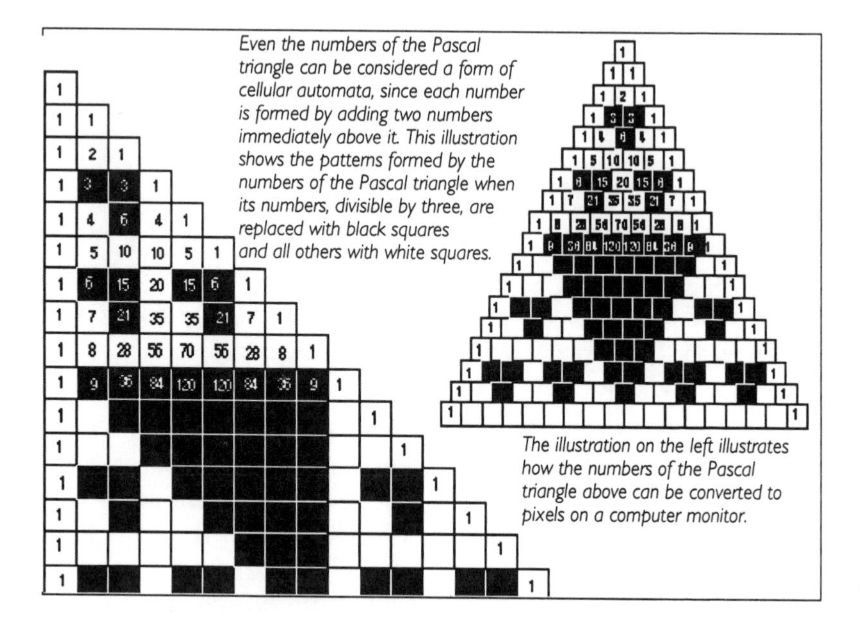
Even the numbers of the Pascal triangle can be considered a form of cellular automata, since each number is formed by adding two numbers immediately above it. This illustration shows the patterns formed by the numbers of the Pascal triangle when its numbers, divisible by three, are replaced with black squares and all others with white squares.

The illustration on the left illustrates how the numbers of the Pascal triangle above can be converted to pixels on a computer monitor.

discrete mathematics and advances in computer technologies are expanding the study of cellular automata and their use as a modeling and simulation tool. The patterns and simulated results that these artificial life forms produce have direct links to many areas of our daily lives. They are being used to explore and explain biological systems such as bacterial growth, spread of epidemics, and the growth of filamentous plants found in blue green algae. Cellular automata are used in the simulations of ecological systems, social structures, prey competition, forest fires, turbulent fluid flow, fluid dynamics and even the working of catalytic converters. These finite mathematical machines need an initial group of cells and a set of rules to make them function. From then on, they begin their evolution which depends on their neighboring cells. From them patterns are spawned which can show properties of self-organizing, self-reproduction, self-similarity which can exhibit systems with chaos, order, or complexity. It is truly astounding to see a finite universe composed of discrete objects describing the riches of our universe, be they physical, chemical, biological, social or philosophical.

Mathematics ...possesses ...supreme beauty...such as only the greatest art can show.

—Bertrand Russell

# mathematics & an art manifesto

In 1936, a now famous group of artists made a public declaration in the *Manifeste Dimensioniste* about the influence mathematics had on their art. *Who were these artists?* Ben Nicholson, Alexander Calder, Vincent Huidobro, Kakabadzé, Kobro, Joan Miró, Moholy-Nagy, Antonio Pedro, Arp, P.A. Birot, Camille Bryen, Robert Delaunay, César Domela, Marcel Duchamp, Wassily Kandinsky, Fred Kann, Kotchar, Nina Negri, Mario Nissim, Fr. Picabia, Prampolini, Prinner, Rathamann, Ch. Sirato, Sonia Delaunay, and Sophie Taeuber Arp signed the manifesto that was drafted by artist Charles Sirato and published in *Revue N+1*.

*Why was the declaration made?* The *Manifeste Dimensioniste* , was prompted by the new ideas coming to light in the mathematics of higher dimensions, relativity and non-Euclidean geometries, and states:

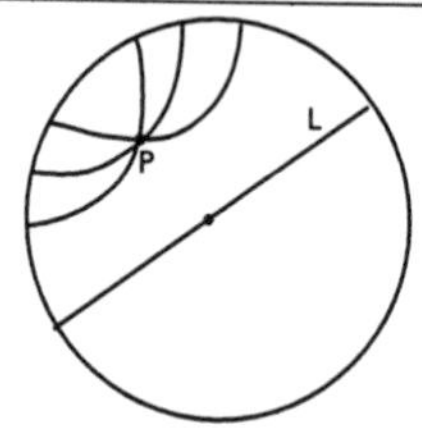


**In Henri Poincaré's hyperbolic geometry model,** *infinitely many lines can be drawn through P. The lines shown passing through P are considered parallel to L since they do not intersect L.*

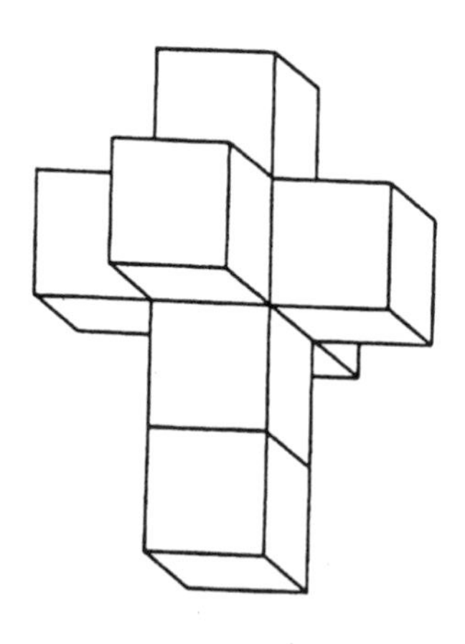

**Higher dimensional objects (such as the hypercube unfolded), non-Euclidean geometries and the dimension of time influenced the art of the artists who formulated this manifesto and many subsequent artists. The hypercube unfolded appears in Salvador Dali's work The Crucifixition (1954), The universe of hyperbolic geometry is found in M.C. Escher's Circle Limit I, woodcut, 1958, and the rubber sheet geometry of topology in his Balcony, 1945.**

*Animated by a new conception of the world, the arts in a collective fermentation... have begun to stir. And each of them has evolved with a new dimension. Each of them has found a form of expression inherent in the next higher dimension, objectifying the weighty spiritual consequences of this fundamental change. Thus the constructivist tendency compels:*

***I. Literature** to depart from the line and move in the plane...*

***II. Painting** to leave the plane and occupy space: Painting in space, Constructivism, Spatial Constructions, Multimedia Compositions.*

***III. Sculpture** to abandon closed, immobile, and dead space, that is to say, the three-dimensional space of Euclid, in order to conquer for artistic expression the four-dimensional space of Minkowski.*

A traveler who refuses to pass over a bridge until he has personally tested the soundness of every part of it is not likely to go far; something must be risked, even in mathematics.

—Horace Lamb

# mathematics finds its way around mazes

What's mathematical about a maze? If something is a problem, then it has the potential of becoming mathematical. You may recall the pastime of walking the seven bridges of Königsberg. Most people just considered it an amusing diversion. But a mathematician viewed it as a logical challenge, and in this case launched an entirely new field of mathematics. Mazes also provide logical challenges, both in their designs and in their solutions. True, rats can learn to solve and remember their solutions to mazes. Robots have been designed to do the same. Comput-

*Mia Monroe walks the maze at Grace Cathedral in San Francisco, Califonria.*

ers have been programmed to exhaust all possible solutions and select the most efficient route. Today scientists even use chemical waves[1] and computers in designing a method that speeds up the computer's solution.

The historical and mystical significance of mazes date back thousands of years. Mazes have been used to ward off enemies, incarcerate monsters and people, invoke mysterious or religious powers, provide amusement for the puzzle solver and as ornaments for buildings and gardens. Well known mazes include —the labyrinth of Knossos with its legend of the Minotaur —the elaborate maze prison of ancient Egypt — the turf mazes and shepherd mazes (outlined in stones) that people used for run-

---

[1] Chemical waves move at a constant speed and get around barriers without dissipating, and disappear at dead ends. Utilizing these properties and computer technology, a group of researchers led by Kenneth Showalter of West Virginia University in Morgantown developed a method to find the shortest route of a maze. See *Pathfinding Made Easier by Chemical Waves* by J. Kaiser, *Science News* February 11, 1995.

*A sketch of the maze found in the Abbey of St. Berlin in St. Omer, France.*

ning competitions — the church mazes set in stone in the floors of cathedrals — the many labyrinth designs etched in stone found in so many different places, cultures and spanning thousands of years — the hedge and topiary mazes of the 17th century — present day amusement park mazes. The fascination of these puzzles has not ceased to intrigue. Children and adults are still captivated by the challenge of doing mazes, as illustrated by the plethora of maze books printed. Today, many churches are exploring once again the spiritual and meditative aspects of walking a maze. The maze at Grace Cathedral in San Francisco, California is a striking example.

*A garden maze at the Governor's Palace in Williamsburg, Virginia.*

Fear not! For those of us who do not want to just try to solve a maze by randomly proceeding along its path or who do not have access to computers or trained rats, the French mathematician M. Trémeaux[2] devised a foolproof general

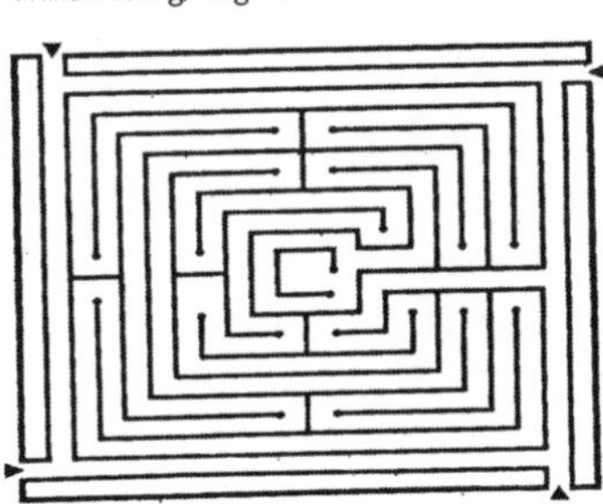

*A garden maze from Architectura Curiosa by G. A. Boeckler, 1664.*

[2]*Récreations Mathématiques*, by E. Lucas, Paris, 1883 vol. I, part iii, pp.44 ff. See also G. Tarry in *Nouvelles Annales de Mathématiques*, 1895, series 3, vol. XIV, pp.187-190.

method that works on any maze.

*Method:*

> *1) As you go through the maze, constantly draw a line to the right.*
>
> *2) Whenever you come to a new juncture take any path you like.*
>
> *3) If on your new path you come to an old juncture or dead end, turn and go back the way you came.*
>
> *4) If returning on an old path you come to an old juncture, take any new path, if there is one. Otherwise take an old path.*
>
> *5) Never enter a path marked on both sides.*

This method does not guarantee the most direct route.

Try your hand at one or all of these mazes.

Another, not surefire way, to solve a maze is:

> *go through the maze always keeping one hand (either the right or the left) in contact with the wall.*

*The famous labyrinth maze of Knossos in Crete.*

# a meter is a meter, is a meter. or is it?

Science is always wrong; it never solves a problem without creating ten more.

—George Bernard Shaw

Many things in mathematics are arbitrary — the symbols used to designate numbers, the base for a place value system, the definition of terms given from arbitrary undefined terms, the unit spacing on a number line, and **units of measure.**

**distance** (length) has been measured in meters, yards, furlongs, leagues, miles, feet, cubits, palms, paces,...

**mass** (weight) has been expressed in such units as scruples, grams, grains, ounces, pounds, kilos,...

**temperature** has been given by Fahrenheit degrees, Celsius degrees, Kelvin degrees ...

**volume** (liquid) has been measured in quarts, liters, cups, ...

**time** measure comes in seconds, hours, days, years, months, hours, nanoseconds,...

Because of the arbitrary nature of designating the units for weights and measures, their evolutionary history has spanned centuries. From ancient times different measuring units and systems were developed in various parts of the world. Among these we find the weights and measures developed by the Babylonians and the Egyptians and adapted by the Greeks and Romans, while Chinese and other Asian cultures designed their own weights and measures.

Some of the earliest measuring instruments were those based on human anatomical parts. For example, the Egyptians defined a *cubit* as the distance from the elbow to the end of the middle finger. The Greeks used the length of the *foot* as their main measuring tool. The Romans used the *pace*, the length of the double step. Naturally no one had trouble having these measuring instruments at hand, but lengths varied depending on whose body was used. Eventually the ancient world developed standards. The Egyptians developed two standard cubits, the *Royal Cubit* which was approximately 20.59" and the *short cubit* about 17.72". The Egyptians even constructed metal bars to correspond to the *Royal* and *short cubits* with subdivisions of *palms* and *digits.*[1] One of the first major contributions to standardization of weights and measures was the Roman law requiring the use of Roman units[2] in all parts of the Roman empire.

---

[1] Each cubit was divided into seven smaller units called *palms*, i.e. the width of one's hand. Each palm was subdivided into four *digits*, namely the four fingers or digits of the hand (excluding the thumb).

[2] The Romans used a base twelve system which divided the *foot* and the *pound* into 12 *unciae* (parts). The unit *libra* was used for weight [the derivation of the abbreviation, lb., for English pound], five feet equaled a *pace*, and a *mille* was 1,000 paces (i.e. 5,000 feet).

But as governing powers changed so did the standards and system of weights and measures. With the fall of the Roman Empire, regions, towns and even guilds adopted their own standards and units. Many English units come from those which originated during Middle Ages—acre, furlong, rod, yard. During this time, especially at fairs, some standards were agreed upon by merchants. A unifying step occurred in 1215 with the Magna Carta, which established standards for measuring grain and wine. Over 600 years later the Weights and Measures Act of 1824 attempted to clean up all the discrepancies that had developed over the centuries. Just as the Roman Empire's system had become the standard, now the British Imperial System spread throughout its colonies. But again discrepancies evolved because newly decreed revisions to the British system did not automatically reach its colonies. Until this time, systems of weights and measures were mainly established by merchants and other laypeople. The first major inroads to a scientific approach to weights and measures took place at the end of the French Revolution. In the late 18th century, the French Academy of Science was asked to devise a new

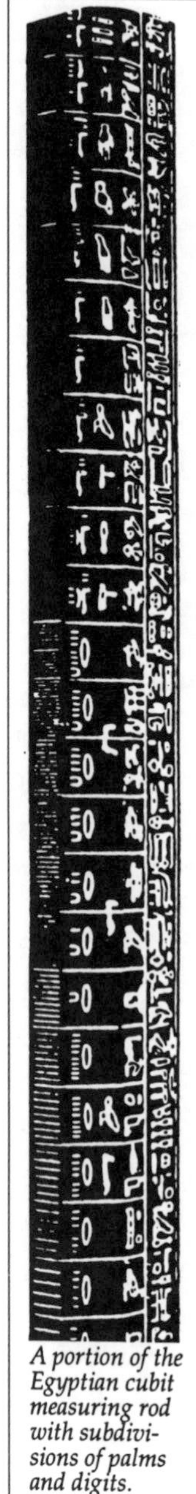

*A portion of the Egyptian cubit measuring rod with subdivisions of palms and digits.*

system to put an end to the mishmash of units being used at that time. In 1840 the metric system became mandatory in France.

The metric system is a base-10 system. The *meter* is its basic unit for measuring *distance.* It was defined as 1/10,000,000 of the length of the meridian from the North pole to the equator and passing through Paris. Other units were defined in terms of the meter. For measuring weight, a *gram* was defined as the mass of one *cubic centimeter* of water at the temperature 4° C, its maximum density. To measure *volume,* the *liter* was defined as the *cubic decimeter.* To regulate and maintain the metric standards, the Metric Convention of 1875 established the International Bureau of Weights and Measures (IBWM) at Sèvres, France. The global scientific world along with most countries[3] have since adopted the metric system as their standard. The IBWM expanded the metric system in 1960 by creating Le Système International d'Unités (SI). Today's standards for weights and measures are no longer governed by empires or merchants, but by scientific needs. With more scientific experiments requiring added accuracy and with new emerging fields such as nanotechnolgy,

| kilo- | hecto- | deka- | meter<br>gram<br>liter | deci- | centi- | milli- |
|---|---|---|---|---|---|---|
| 1000 | 100 | 10 | 1 | .1 | .01 | .001 |

*See appendix for additonal prefixes used in the Systeme International*

[3] The US Federal Trade Act of 1988 sought conversion to the metric system, but as of yet the metric system is not used in the daily activities of the citizens.

scientists are again revising the standards. So what has happened to some of the old metric standards?

• *The length of a* **meter** *is no longer determined by the distance between two marks on a platinum-iridium*[4] *bar stored in a vault at the IBWM in Sèrves, France. Now the meter is the distance light travels in 1/299,792,458 seconds.*

• *A* ***second*** *is no longer what most of us know it as, 1/60 of a minute. A second has been redefined as the time it takes a cesium atom excited by microwave radiation to vibrate 9,192,631,770.*

• *The* ***boiling point of water*** *was 100° Celsius, but it to has been revised to 99.97°C by using molecular motion rather than properties of water.*

So what's happened to the ***kilogram***? Its weight seems to be the next one slated for revision. When agreed upon it will no longer be preserved by a platinum-iridium cylinder in Sèvres, France. Scientists are working on deriving a universally agreed upon method for determining the weight of a kilogram as a particular number of atoms by using atomic weight, electronic measurements and Planck's constant, which brings in quantum physics.The National Institute of Standards and Technology in Maryland has weighed a kilogram against an electromagnetic force and are looking for methods to reach the required accuracy for the Planck constant. Another method being explored is counting the number of atoms in a kilogram of silicon. With such methods being perfected, the atomic kilogram is in the near future.

Will this be the end to changing the values of the units of measures? History is full of examples witnessing the evolutionary change of measuring units, so don't expect the process to end.

---

[4] This alloy is highly heat and corrosion resistant and chemically nonreactive, which made it ideal for preserving the defined IBWM standards of length and weight since its given size and weight was essentially retained under a spectrum of conditions. But it too, no matter how slowly, does change, and thereby both its weight and length are minutely affected over time. With scientists requiring more and more accurate and precise measurements, the platinum-iridium standard had to be changed.

I saw...a quantity passing through infinity and changing its signs from plus to minus. I saw exactly how it happened...but it was after dinner, and let it go.

—Sir Winston Churchill

# molecular computers

Imagine a computer that is a thousand times faster than a supercomputer — performing a trillion operations per second. Imagine a computer using one-billionth the energy used by conventional computers and able to store a trillion times more information in the same allotted space. Where can one find such a computer? Such computers reside in the cells of our bodies, in DNA. They are molecular computers. Although molecular computers have not as yet been commercially built their potential was demonstrated in November of 1994 by computer scientist Leonard M. Adleman of the University of Southern California. Adleman wanted to test the feasibility of using a molecular comput-

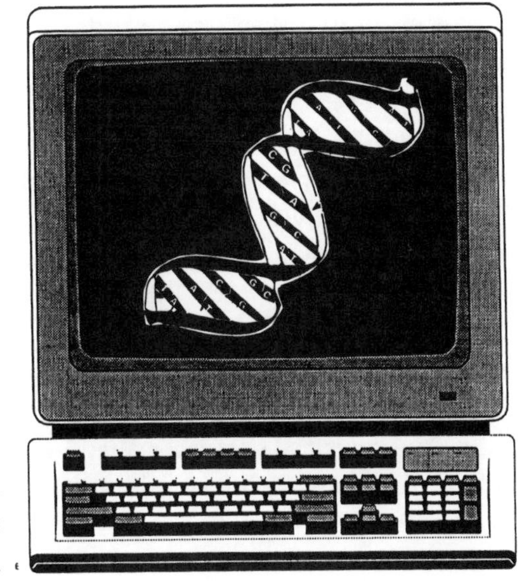

er to solve a mathematical problem, but he had a number of obstacles to tackle. What type of problem to test? How to use the workings of DNA to come up with a way to program the problem and also decipher the solution? Coming up with such a problem to solve, then figuring out how to implement its solution was quite a feat. Adleman went with a problem which may be initially reminiscent of the Königsberg bridge problem, but is in fact a Hamiltonian path problem[1]. He decided to choose a problem that has only been solved by testing all possibilities using conventional computers. The problem was to find the shortest path that links seven points, one of which is designated as a starting point and one as an ending point. This problem is analogous to a traveler wanting to find the shortest route from San Francisco to Rome while stopping along the way at each of five other designated cities only once. How do you tell DNA to solve such a problem?

Adleman assigned each city a DNA first and last name. By using the four nucleotides of DNA, namely A, C, T, and G, each name consisted of

---

[1] In 1857 Sir William Rowan Hamilton created a game called *Around the World.* A dodecahedron was used with each of its 20 vertices named after a city. The object was to travel along the edges of the dodecahedron and visit each of the cities just once. A path that passes through each vertex exactly once has come to be called a *Hamiltonian path.* If the path ends with its starting point , it is known as a *Hamiltonian circuit.* In the Königsberg bridge type problem (called an Eulerian circuit), the path must go over each of the brigdes exactly once, while each vertex may be passed through as often as necessary. The Hamiltonian path does not have to travel every edge, but each vertex must be visited only once.

a sequence of 10 arbitrarily selected letters for the first name and 10 for the last name. For example, San Francisco's first and last names could be arbitrarily assigned the nucleotides[2] in this order ATTGACTCGC and CTGGAACGTG. Then names for each flight between two cities utilize the first and last names of cities. For example, a flight going from San Francisco to Miami would use the last DNA name of San Francisco and the first DNA name of Miami. In addition, since these nucleotides have a tendency to bond in a specific way with their complements (A with T and C with G), Adleman then designated the complementary names for each city to assist in bonding. The names for each city, each flight and each complementary name were formed into DNA strands. Then all these DNA strands were mixed in a test tube. Immediately each DNA flight strand began to link to its cities and trillions of long chains of DNA began to form. The molecular computer was at work. Within a moment all possible routes were formed in the molecular solution. One of these chains had the right originating city and the right destination city plus all the other five cities, and was the shortest in length. Using established laboratory techniques for molecular biology, Adleman was able to isolate in the solution the desired DNA chain in about a week. Thus, he created a way to utilize the properties of DNA molecular bonding to solve a mathematical problem. His design used the four basic units of DNA to translate and solve a problem somewhat analogous to the working of traditional computer and its use of 0's and 1's.

The number and types of computers that have been invented over the centuries is a tribute to human ingenuity and curiosity. Among these we have the abacus, the quipu, the analog, the digital, the microchip. Molecular computers, also known as DNA or biological computers, are not far-fetched, especially when one considers how effortlessly DNA and the brain regulate and monitor all the inifinite functions of the body.

[2] Nucleic acids are formed from nucleotides, which are made from a sugar, a phosphate and a base (there are five bases which are referred to by the symbols A for adeline, C for cytosine, G for guanine, T for thymine, U for uracil). In DNA, only four bases A, C, G and T appear. Each base can be paired with only one other complementary base — A with T and C with G — and in DNA two strings of nucleic acids join together at their bases and form a double helix.

Everything in nature adheres to the cone, the cylinder, and the cube.

—Paul Cezanne

# *a mathematical look at the* art of Cezanne

Trying to find the mathematical aspects of works of art can sometimes be forced or superficial. For ages, artists have often analyzed art technically and mathematically in order to help them achieve a particular goal. For example, the hallmark of Renaissance art is the realism and accuracy of the works. To this end many of the artists of this period analyzed their work and subjects in minute detail, and their studies focused on attaining an understanding of perspective using projective geometry. The location of an object in space and the use of lines of light and form dictated the perspective the artist wanted to achieve. To enhance understanding, even the human body was dissected both literally and figuratively by studying its anatomy and every possible symmetry[1] it possessed. In

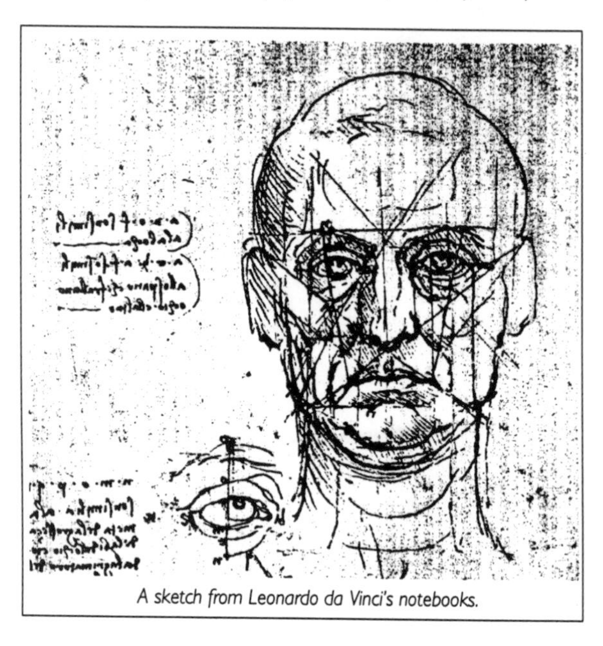

*A sketch from Leonardo da Vinci's notebooks.*

essence, the human body was viewed and broken down into basic geometric forms to gain insight on how to recreate it precisely. Hence, the Renaissance masters focused on the structure and realism of their works, and their technical analysis provided the artist new insights and freedom in the formulation of works. The same is true with the work of M.C. Escher, who struggled with the formation of tessellations until he unraveled the mathematics behind their formation. A self-analysis of the work of Paul Cezanne (1839-1906) does not exist, yet many art critics have scrutinized and compartmentalized his work. "When one knows

---

[1] Leonardo da Vinci's *Trattatto* and Albrecht Dürer's *Treatise on Human Proportion* are examples of such studies.

*Cézanne's Still life with Fruit Basket (1888-1890). Musée D'Orsay. Paris, France.*

how to render the cone, the cube, the cylinder, the sphere in their form and their planes, one ought to know how to paint," sums up Cezanne's analysis of art. But it is mathematically fascinating to look at how Cezanne viewed space.

Cezanne's works do not focus on conveying realism, but focus instead on a totally different aspect of art. He is regarded as the organizer of space. What is the difference between Renaissance space and Cezanne space? In Euclidean geometry, space is the set of all points. When a Renaissance artist placed an object in space (on the canvas) accompanied by relevant scenery in perspective, space in this sense was incidental, just the means to showcase the subject being drawn. Cezanne presents an entirely different notion about space. In his work, space is as integral a part as the subject of the work. To him the subject occupies a set of points of space and the complement of this set (i.e. all other points of space) is also a theme of his work. He develops space with its own mass (structure) while the object's structure is somewhat understated so that it merges with the space around it. Thus space and the object are a unit forming the entire space—in this way forming the entire work. In addition to this use of space, many of his works illustrate transformation of forms. The line takes on a new sense. It is no longer the ideal Euclidean line—straight, distinct and continuous. Instead his lines often appear broken, not straight and/or fading from the drawing. He sometimes distorts perspective, similar to what one expects to find in a non-Euclidean world of elliptic geometry. His *Still Life with Fruit Basket* illustrates:

— how the "sacred" Euclidean line is no longer straight and fades away ( note the line formed by the edge of the front table)

— how the depth of space is distorted yet its infinite nature is conveyed by having parts of the furniture cut-off abruptly

— how multiple points of perspective are created (the plane of the handle of the basket implies one perspective, the handle of the pot points another direction, the angle of the wall and the location of the table show other perspectives) in space and create a wide periphery of vision. And, as mentioned, all objects in this space seem to possess equal importance—the fruit, basket, pots, chair, tables, floor— all merge into a single unit.

Cezanne is well known for other artistic techniques, especially phenomenal use of color to create dimension, but his representation and organization of space and its depth are especially mathematically significant.

# *where am i?* mathematics and the global positioning system

All science requires mathematics.

—Roger Bacon

On March 29, 1996 President Clinton signed a Presidential Decision Directive allowing the civilian and commercial use of the Global Positioning System (GPS). At the signing, Transportation Secretary Federico Peña pointed out that "Today, not many Americans know what GPS is. Five years from now, they won't know how they ever lived without it."

GPS, developed and operated by the U.S. Department of Defense, was initiated in 1978, and as of July 17, 1995 it reached Full Operational Capability. It is a radionavigation system that is satellite-based. 24 satellites are positioned in six circular orbits above the earth. They are arranged so that at any time a minimum of six satellites are always in view to any user around the world. These satellites continually broadcast their position and time to the Ground Control Segment which processes and updates the information into navigational messages for each satellite. The satellites function as precise reference points for anyone capable of receiving their signals. Users receiving this information can determine their precise location, velocity and time.

*How does the GPS work?* The Global Positioning System[1] (GPS) may use advanced technologies, such as satellites, advanced telecommunication systems, and atomic clocks synchronized to the nanosecond; but to find the exact location of a sailboat, a truck or anyone or thing equipped with a transceiver, it still relies on the ancient method of *triangulation.* Triangulation is what ancient seafarers and travelers used to navigate themselves over land and sea. It is based on geometric and trigonometric concepts. For example, if the measurements of only one side and two angles of a triangle are known, the other sides and angle can be determined.[2]

Today, it is all done by means of measuring the user's distance from a group of satellites in space. To communicate with the satellites, a vehicle must be equipped with a transceiver, a devise which transmits & receives signals from the GPS satellites.

To determine the location of a lost vehicle a minimum of three satellites

---

[1] GPS provides a Standard Positioning Service (SPS) for the general public use and an encoded Precise Positioning Service (PPS) mainly for use by the Department of Defense. For security reasons the SPS is less accurate.

[2] Whenever two angles of a triangle are known the third one can be easily computed because from Euclidean geometry we know the three angles of a triangle total 180°. Now we know the measurements of the triangle's three angles and one of its side. To find either of the unknown sides, one can use the law of sines of trigonometry.

and their respective distances from the object are needed. Visualize each satellite as the center of an imaginary sphere. Recall a sphere is the set of all points in space equidistant from a given point, called its center.

Each satellite is equipped with four atomic clocks (one is the primary clock , one is a backup and the other two keep the first two synchronized). At every instance, each satellite has three coordinates that identify their position — altitude, longitude, and latitude. To determine the car's location, each satellite first determines its distance from the car, using the formula, *speed•time=distance* . Here's where the atomic clock is important. When the satellite receives a signal , that signal has the time it was sent recorded in it, and the satellite notes the time the signal was received. Then the GPS software computes the difference between the two times (the send time and the receive time). For example, this may be 0.09 seconds. The speed of the signal is the speed of light, approximately 186,000 miles per second. Thus the car's distance from this satellite equals 186,000•0.09 (i.e. speed x time), which is 16,740 miles. In the 3-dimensional view this means the car at that moment could be at any point in space that was 16,740 miles away. Each of the three satellites' distances from the car is calculated in this way.

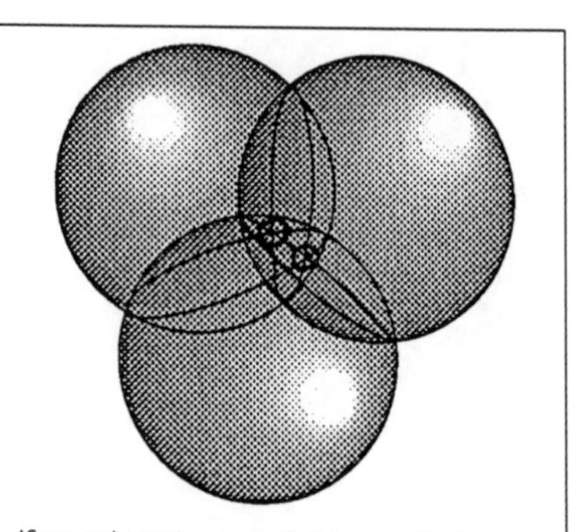

*If two spheres intersect, their intersection is a set of points on a circle. If a third sphere intersects these two spheres, the three spheres now have only two points in common. Add a fourth intersecting sphere, and only one point will be in common. GPS systems use three or more satellites.*

Now, visualize an imaginary sphere around each satellite, with the satellite as the center of the sphere. The imaginary sphere's points represents all the possible points in space that the car can be located with respect to a particular satellite, since the radius of each sphere is the car's respective distance from each satellite. These three spheres can intersect in only two points. One of those two points is the car's location, while the other is a point probably off the Earth. The time it takes the GPS software to complete these calculations can be as little as a few seconds.

GPS can be incorporated in traffic management centers, buses, trains and emergency response systems. It will be the foundation for guidance systems in aviation, highway, maritime, trucking, rail and mass transit. Now it even provides guidance on an individual level— from wilderness exploration to location guidance for the blind.

Secretary Peña pointed out in his address that "The birth of this new industry will completely transform the way we live from driving to work, to delivering a package, to responding to an emergency. GPS will have profound effects on everyday life, saving people time, cutting costs and giving people mobility and choices that they never had before."

# mathematics in literature

I shall persevere until I find something that is certain — or at least, until I find for certain that nothing is certain.

—René Descartes

When most people think of mathematics, numbers, complicated equations or diagrams come to mind. Mathematics is always linked to the sciences. When we think of mathematics books we think of non-fiction, even though mathematics itself is predominantly fiction. Its ideas and concepts are all figments of the imagination. In Euclidean geometry a point, a line, a plane are all imaginary. Their ideal forms cannot exist in our world. The same is true of squares, triangles or circles — all are imaginary. Of course, we can see a representation of a square on paper, but it would be physically impossible to draw four perfect 90° angles and four sides precisely the same length. We know how mathematicians arbitrarily divided a circle into 360°, and the location of the origin on a line thereby transforming it into a number line is purely arbitrary. The objects of mathematics seem real to us, especially when we use them to describe things around us. The negative integers seem real when applied to temperatures or budget deficits. Fractals used in cinematography seem real when they become landscape. But once again these are things of fiction. We have no problem accepting far out mathematical ideas occurring in such works as *Star Trek, Contact* and *Star Wars,* but how many think of mathematical ideas in relation to literature? Yet, mathematical concepts have indeed had roles in many works of fiction.

Cold statistics, probability and complexity theory play major roles in the Robert Coates short story *The Law.* Here we read that *"The hours from seven till midnight are normally quiet ones on the bridge. But on that night it was as if all the motorists in the city ... had conspired to upset tradition."* This is a story of what happens when things don't go as expected or predicted —unexplained traffic jams, customers flocking to a particular store one day while leaving it deserted the next day. Coates talks about the law of averages, but in view of today's mathematics it is also a wonderful example of complexity theory at work and the tenuous balancing act between order and chaos.

In the novel *Einstein's Dreams*, Alan Lightman does a masterful job with a fictional account of the thoughts and dreams of Albert Einstein while he was formulating his theories of time and relativity. Swept up in the book's various philosophies, you are unaware that you are learning about Einstein's theories, and that these theories are the underlying plot line. In one chapter a dream deals with the supposition that time moves in a circle and the story ends with the paragraph *"In the dead of the night these cursed citizens wrestle with their bedsheets, unable to rest, stricken with the knowledge that they cannot change a single action, a single gesture. Their mistakes will be repeated precisely in this life as in the life before. And it is these double unfortunates who give the only sign that time is a circle.*

*For in each town, late at night, the vacant streets and balconies fill up with their moans."*

Many of Umberto Eco's novels, such as *The Name of the Rose, Foucault's Pendulum* and *The Island of the Day Before* have mathematical ideas woven into the story, adding an added level of realism to the plot. *In The Name of the Rose* codes and ciphers play an important role. The first lines in *Foucault's Pendulum* are

> *"That was when I saw the Pendulum.*
> *The sphere, hanging from a long wire set into the ceiling of the choir, swayed back and forth with isochronal majesty.*
> *I knew—but anyone could have sensed it in the magic of that serene breathing—that the period was governed by the square root of the length of the wire and by π,..."*

Such mathematical phrases, besides adding a level of credence to his work, cast a veil of mystery.

In Jorge Luis Borges' short story *The Book of Sand,* the main character acquires an incredible book which has no beginning and no end. *"It can't be, but it is. The number of pages in this book is no more or less than infinite. None is the first page, none is the last. I don't know why they're numbered in this arbitrary way. Perhaps to suggest that the terms of an infinite series admit any number."* The infinite book and what it implies occupies every thought and moment of the main character. Realizing the book would consume him, he knows he must get rid of it. Naturally he could not tear it to pieces, for the Earth does not have enough room to store the trashed book. This story is about more than just infinity. Even though it was written in the early 1900s, when we read it today we can't help thinking of reiteration processes and fractals and the way a fractal can grow and change before our eyes.

In the last few years, Tom Stoppard's play *Arcadia* took theatergoers by storm. Set in both the 1800s and the present, the heroine(Thomasina) of the past is a precocious young woman who is exploring how mathematics can be used to describe the world around us. In one seen she challenges her teacher, when she says:

> *"Each week I plot your equations dot for dot, xs against ys in all manner of algebraical relation, and every week they draw themselves as commonplace geometry, as if the world of forms were nothing but arcs and angles. God's truth Septimus, if there is an equation for a curve like a bell, there must be an equation for one like a bluebell, and if a bluebell, why not a rose? Do we believe nature is written in numbers?"*

She records some revolutionary discoveries, which do not surface until the present. Her present day counterpart is a mathematician working on the theory of complexity and chaos— here again these ideas are pervading elements of the plot line.

Lastly, consider Ray Bradbury's short story *Sound of Thunder,* in which the characters take a recreational time travel trip and fall victims to its consequences. Here a butterfly's demise *"upsets balances and knocks down the line of small dominoes and then bigger dominoes and then gigantic dominoes, all down the years across Time...It couldn't change things. Killing one butterfly couldn't be that important! Could it?"* Is this story a precursor to the *butterfly effect*[1] described in chaos theory where simple minute differences in the initial input could result in enormous differences in the outcomes?

Mathematics has also figured in many movie screenplays. For example, consider the following movies whose screen plays deal with mathematical themes and ideas: *Good Will Hunting* received an Oscar for *Best Original Screenplay* in 1997. The 1998 movie *Pi*, is a story about a genius using mathmatics to explain the bigger picture of life and its meaning. In the 1980 comedy with Jill Clayburgh, *It's My Turn*, the *snake lemma* is proven. The negative/positive sign switch causes big problems for Dustin Hoffman in the 1971 movie *Straw Dogs*. *Conceiving Ada* produced in 1998 introduces us to Ada Lovelace and her world, while the 1983 *A Hill on the Dark Side of the Moon* deals with Sonya Kovalevsky and other mathematicians such as Karl Weierstrass and Gösta Mittag-Leffler.

These and many more works are entertaining, but even more important they are springboards for stimulating one's thoughts and exploring mathematical ideas. One thing always seems to lead to another.

---

[1] In Chaos Theory the phrase *the butterfly effect* is used in connection to chaotic weather changes. An analogy is drawn on how the fluttering of the wings of a butterfly in one part of the world could start a chain reaction of small air turbulences resulting in a full scale hurricane in another part of the world.

Many who have never had an opportunity of knowing much about mathematics confuse it with arithmetic, and consider it an arid science. In reality, however, it is a science which requires a great amount of imagination.

—Sonya Kovalevsky

# crime & complexity

*"The hipbone is connected to the backbone, the backbone is connected to the..."*

Everything we do, see, or hear in some remote often unknown way is connected to something else. We put salt on our food, our blood pressure often goes up. The winter rains feed the plants and rivers, which affect the fish in the streams which affect the fishermen's catch affecting the price of fish, our diet, how much we spend....

The chemicals in the air, in the soil, in the sea may enter our food chain and trigger the genes of certain people to act adversely and thereby set off a chain reaction altering the normal functions of their bodies.

As John Muir said, "When we try to pick out anything by itself we find it hitched to everything else in the universe."

At the root of explaining this connectivity is a language, complete with its own alphabet and grammar (operations). A language that has been evolving for thousands of years in all parts of the world, and is used by all peoples of the world. A language that has touched all people, in all places. A language that is becoming more complex as our lives become more complicated. **The language is mathematics.** Over the centuries mathematicians have tried to use mathematics to explain the workings of the universe. Among the various efforts, we have the Pythagoreans who tried to explain the universe in terms of the whole numbers and their ratios. Time has shown the universe is more complex than that. Some scientists felt the universe could be described using the Platonic solids as illustrated by the model of Johann Kepler, but we have learned our solar system is more complicated than that. Leopold Knonecker did not feel transfinite numbers were of use or necessary, yet now they are indispensable to mathematics. And now, as our probes and telescopes reach into deep space transmitting new data and photographs, scientists are groping excitedly and frantically for explanations for the discoveries of such things as black holes, a fountain of antimatter, the birth of stars, orientation of the universe.

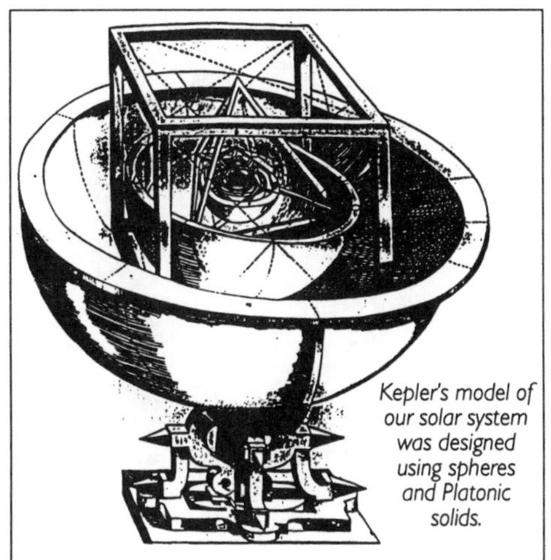
*Kepler's model of our solar system was designed using spheres and Platonic solids.*

Time and again, we discover the universe is more complicated than the present status of mathematics and sciences can explain. What happens? Mathematics expands. Mathematicians invent new mathematics to explain new ideas, or discover applications for ideas previously considered intriguing oddities. For example, when mathematicians first came across

problems that required imaginary numbers, such as $x^2 = -1$, they chose to regard such problems as having no solution. Then in the mid-1500s Gerolamo Cardano proposed a problem that required the solution of the equation $x^2-10x+40=0$. Instead of being satisfied with saying it had no real number solutions, he solved it finding it required a number that was the negative square root of 15, which was non-existent at that time. Complex numbers did not receive acceptance until the 1700s. Today complex numbers are known as two-dimensional numbers, and are used in a variety of ways such as to describe patterns of flows of hydrodynamics, electric currents and even the shapes of airfoils. Quaternions, four-dimensional numbers, were invented by Sir William Hamilton in 1843, and now they are used in communicating graphics information on computers by giving description of rotations in three dimensions.

Here in the midst we find complexity theory trying to deal with very complicated and connected universes. Complexity is the science that explores complex systems of the universe whose infinite facets many people feel defies explanation.

Natural phenomena are not the only areas where complexity theory is used. Complexity can apply to any non-linear systems, systems that do not exhibit direct cause and effect characteristics. These systems have a balancing or tipping point which when disturbed can result in chaos. Weather is a prime example of a complex system, as is the spread of disease, the balance of power between countries, the economic stability of a country. Each of these can be affected by almost imperceptible changes taking place, which in turn can lead up to the tittering point causing the chaos in Bosnia or the seemingly swift change of power in South Africa.

Imagine a big city where the principles of complexity theory are used to fright crime? Farfetched? New York City's crime problem is an excellent example of a complex system at work. We are familiar with stories about rampant drugs and violence in the streets of New York City, but in 1995 New York City's crime picture changed dramatically. It was no longer ranked among the top crime centers, but dropped to 136th. Car thefts fell from 150,000 in 1993 to 71,000 in 1995. Burglaries were down from over 200,000 in 1980 to below 75, 000 in 1995. Homicides were reduced by nearly 50%. Violent crimes reported from every precinct had double-digit decreases. In fact, the most violent crime ridden neighborhoods had the greatest decrease in crime. What influenced this about face? How did these changes occur? No one thing can be given credit, but notably, the New York City Police Department, under the leadership of Police Commissioner William J. Bratton, consistently made changes, sometimes imperceptible positive changes, toward improving the crime situation. Among these changes were more coordination between divisions, streamlining internal procedures, more accountability from precinct commanders, more arrests for gun possession, ongoing removal of graffiti, dispersing of congregating youths, increase in safety checks for drunk drivers and stolen cars. Changes were made consistently and persistently. In crime areas, where people had feared to venture out, they began to reclaim their world. This in itself acted as a factor for change. One thing led to another. Each action did not necessarily produce a result. When positive change came, it was not gradual, but seemed to happen all of a sudden. Just as suddenly, the balancing act can reverse if negative changes begin to mount causing the seesaw to tip the other way — *the butterfly effect* [1] in the world of crime. Today, police brutality tied to racism and bigotry has escalated in New York as well as in Connecticut, Maryland and California. Will these and other negative changes alter the precarious balance of New York City's crime situation?

Life and human behavior is a complex system, and mathematics can be used to describe, influence and possibly predict such behavior.

---

[1] The analogy of a butterfly is used in chaos theory to describe changes in a complex system. For example, in one part of the world a butterfly flapping its wings could start small air turbulence that could multiply and result in a full scale hurricane in another part of the world. In other words, small almost imperceptible changes in initial conditions can have enormous consequences.

Time present and time past
Are both present in time future,
And time future contained in time past.
If all time is eternally present
All time is unredeemable.
What might have been is an abstraction
Remaining a perpetual possibility.
Only in a world of speculation.
What might have been and what has been
Point to one end, which is always present.

— T.S. Eliot

# a mathematical look at time

Over the centuries time has tantalized and perplexed the imaginations of philosophers, poets, artists and scientists. Today time still remains an illusive concept. *What is time?* Does it always move forward? Does it move at all? Is time a figment of the imagination devised to help order and organize our lives? We mark time in so many ways. Humans have invented a host of mechanical devises to keep track of time. They've marked the seasons with Stonehenge, the hours with sundials and water-clocks, the seconds with pendulums, and minute fractions of a second with atomic clocks. We would like to be able turn the clock back or speed it forward, but only in the realm of science fiction has this been possible. Physicists and mathematicians entertain these notions, but thus far with much frustration.

Through the years, time has been linked to motion— to the movement of planets, stars, the moon, comets. The ancients relied on the moving and changing seasons to determine when to plant and harvest. They noticed the cycles of plants. Now we know that all things are in constant flux or change. All things, animate and inanimate, have timepieces within them on a molecular level. A certain butterfly's entire life cycle is but a day, while carbon-14's half- life is 5,730 years. Even the Sun will undergo a major metamorphosis in billions of years. *The key to the passage of time is change.* Without observing change, time would appear to stand still. Change implies motion— be it the motion of the Sun across the sky or the molecular motion of an electron. To measure change or motion, units of time and distance come into play. In algebra we learn how time, speed and distance are connected by a simple mathematical formula ($d=s \cdot t$), which seems to work well for our daily lives. Yet on a universal scope, time may not behave as we are accustomed. We learn in modern physics that space and time are linked. Since Einstein's theories, space and time are no longer considered absolutes. Time is variable as are space, matter and everything in the universe except for the speed of light. Einstein's theories assume the speed of light absolute. It is the speed limit of the universe. The time or the aging of an object slows down as it approaches the speed of light; its dimensions shrink and its mass increases.

The story of time goes hand in hand with the story of the universe. Various theories have been formulated to describe the origin, the ending, the present state of the universe. Centuries have been peppered with many diverse ideas about its workings. Starting with Aristotle's ideas of motion, Galileo's laws of uniform acceleration, Newton's use of time in his definition of momentum[1] and laws of gravity— time was considered

---

[1]Newton formulated that uniform motion was relative, but refused to accept that time was not absolute.

absolute. A minute was the same whoever observed or experienced it. But Einstein's special and general theories of relativity changed the role of time. Yet even Einstein's views of time become questionable in the ultra small world of particles. Here the mathematics of probability[2] and statistics play important roles. Quantum theory deals with the universe of particles and subparticles. So many fascinating and incredible things exist at this level. Here photons and neutrinos travel at the speed of light[3]. Here particles can exist in two places at the same moment. Here scientists theorize about the existence of tachyons, which can travel at speeds *faster than light* and possess mass equal to an imaginary number with imaginary dimensions while having a real value for time. How can this even be considered? Because these values exist in equations which

---

[2] Einstein was critical of the use of probability in quantum theory.

[3] Photons and neutrinos are the only particles known to travel at the speed of light.

convert their imaginary values to real numbers. We are told that most of the matter (90%) composing the universe is invisible. It is called dark matter. Some scientists believe that dark matter is composed of exotic matter. Exotic matter has been labeled with such names as WIMPS (weakly interacting massive particles), MACHOS (massive compact halo objects the size of Jupiter). How do we know it is there? Mathematical computations reveal that the matter that is visible in the universe is not sufficient to account for the gravitational forces present in the universe. This universe is probabilistic — all probable situations can exist as separate (parallel) universes which constantly change as each new possibility occurs. Each parallel universe has its own past and its own present. If one travels back in time and affects the outcome of the future, a universe with this new future now exists independent of the original one the traveler departed. In the quantum world, we also find mini wormholes appearing in what is called quantum foam, although we have no evidence or explanation for the existence of wormholes on a macrocosmic level. If wormholes could occur on a macro level, a traveler could traverse gigantic distances through space by simply passing through the tunnel connecting both mouths of the wormhole. In addition to traveling great distances, one could possibly time travel via the wormholes. But with the possibility of travel in time comes a host of paradoxes that leave scientists perplexed. How to reconcile Einstein's work with quantum physics? Should not a theory that applies to one work for the other? Can we come up with one explanation, one unifying theory to explain the workings of the entire universe(s)? Will a string theory with its 10 or 26 dimensions provide the answers and the Theory of Everything? The mathematics needed to deal with these ideas is extraordinarily difficult.

If we could forget about time, and ignore its paradoxes, what would we have? Picture the universe moving backwards and forwards simultaneously? These ideas are difficult for our minds to reconcile since the clocks we have invented only seem to move in one direction. With so little time in the human life span when compared to the universe, it's no wonder we want to keep track of every moment. Scientists seek to know the moment of our beginning. Does it include the big bang? It is believed that the universe did not move into existing space, but instead the big bang created matter and space. It began at a point of singularity where an object has 0-dimension and mass of infinite density. At points of singularity, the laws of physics break down and there is no mathematics capable of describing them. Will there be a big crunch[4]? Are the big bang and the big crunch cyclical? Some scientists believe that there was not a big bang but a series of still evolving minibangs. All these questions are theories scientists are exploring in their quest for a cosmic genealogy. No doubt about it, time and mathematics are major players in the exploration and explanation of the universe and its origin.

---

[4]The big crunch is the reverse of the big bang. Here everything will revert to a point of singularity. Some scientists are theorizing that the big bang and big crunch are cyclical events.

# the universe

*what's with it anyway?*

Mathematics ... is limitless as that space which it finds too narrow for its aspirations; its possibilities are infinite as the worlds which are forever crowding in and multiplying upon the astronomer's gaze.
—James J. Sylvester

With scientists able to delve deeper and deeper into space and space probes gathering more and more information, determining the shape of the universe is in a state of flux. It is believed that the future of the universe, whether it will expand itself into oblivion, recede and collapse into itself, or remain static, depends on the curvature of space. What affects curvature? The curvature of the universe depends on the density of its matter, and it is believed there exists a critical level of mass (called critical density) which establishes its curvature. How does matter cause curvature? Consider a large mass, such as our sun. Visualize the sun as a huge ball resting on a rubber sheet. A smaller mass near the sun, such as the earth, would incline toward the sun's depression. This is often referred to as the "pull" of gravity. The sun's mass in essence has warped or curved the space around since space is not flat, the properties Euclidean geometry do not hold. Even the Earth's orbit can be explained as a journey along a non-Euclidean curved line[1], and it's along such a curved space that light itself bends. Until the 19th century, Euclidean geometry was the only geometric system, but serious challenges to Euclid's parallel postulate materialized. Though these challenges did not disprove Euclid, they spawned non-Euclidean geometries —geometries that altered Euclid's parallel postulate.

Different shapes have different curvatures. For example, the curvature of a sphere is defined as 1/(radius). A small sphere has greater curvature than a larger one. For a sphere, the smaller the radius the larger the curvature. Consider the curvature of an egg. It is smaller at the fat end and larger at the narrow end. Concave surfaces are defined as having positive curvature, as in the case of a sphere. Convex surfaces, like the saddle or a pseudosphere, are defined as having negative curvature. Some surfaces have both positive and negative curvatures depending on the region considered, while a flat plane has zero curvature. The properties of geometric objects can change depending opon which

---

[1] In his lecture of 1854 Georg F.B. Riemann (1826-1866) established the existence of a non-Euclidean geometry referred to as spherical or elliptic geometry. Here, finite lines without any beginning or end were defined as the great circles of a sphere. In this geometry infinitely many lines can pass through two pints (e.g. the longitude lines passing through the poles of the sphere). In his lecture Reimann discussed a 3-dimensional universe that is warped in the 4th dimension. In the early 1900s Albert Einstein developed his general theory of relativity ( or gravitation) in which he connected gravity with the curvature of the 4-dimensional non-Euclidean space.

In 1829 Nicholai Lobachevsky (1793-1856) and Johann Bolyai (1802-1860) in their efforts to prove Euclid's Parallel Postulate also discovered a non-Euclidean geometry called hyperbolic geometry. Carl Gauss had also discovered this non-Euclidean geometry, but chose not to make his discovery public, possibly fearing ridicule.

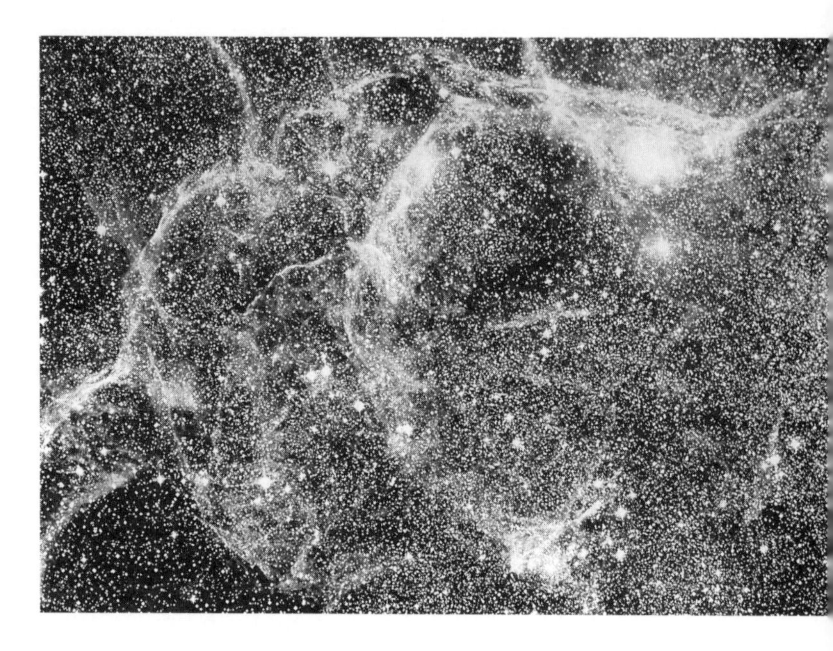

geometry is being considered. For example, on a flat plane, Euclidean properties hold. Here, for example, the shortest distance between two points is a straight line, the angles of a triangle total 180°, parallel lines always remain the same distance apart, while in non-Euclidean geometries these properties are different. In the spherical geometry of Riemann the shortest distance between two points is a curved line, the angles of a triangle total more than 180°, and there are no parallel lines; while in the hyperbolic geometry of Bolyai and Lobachevsky the shortest distance between two points is a curved line, the angles of a triangle total less than 180°, and parallel lines are asymptotic[2]. In spherical geometry, space has positive curvature because a sphere is concave. In hyperbolic geometry, space has negative curvature, as represented by the convex form of a pseudosphere.

---

[2] Asymptotic curve is a curve which continually gets closer and closer to a line (called its asymptote), but never intersects it.

In which of these geometries does our universe exist? What is the curvature of our universe? Scientists believe the critical density determines whether the universe's curvature is positive, negative, or zero. If the matter's density is less than the critical density, the universe has negative curvature and is expanding ad infinitum. Such a universe would be governed by hyperbolic geometry. If, however, the amount of matter is exactly at the critical density, the universe has zero curvature— it is flat and we would be back to Euclid's world. Here the universe would continue to expand, but its rate of expansion would tend toward zero. If the amount of matter exceeds the critical density, the universe has positive curvature, and would eventually turn in on itself, contracting into a sphere and collapsing on itself. Such a universe would be described by spherical geometry.

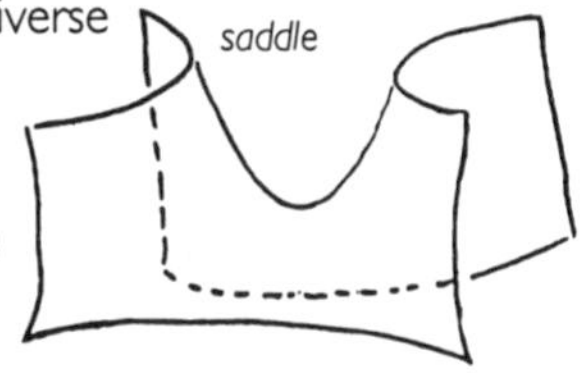

*saddle*

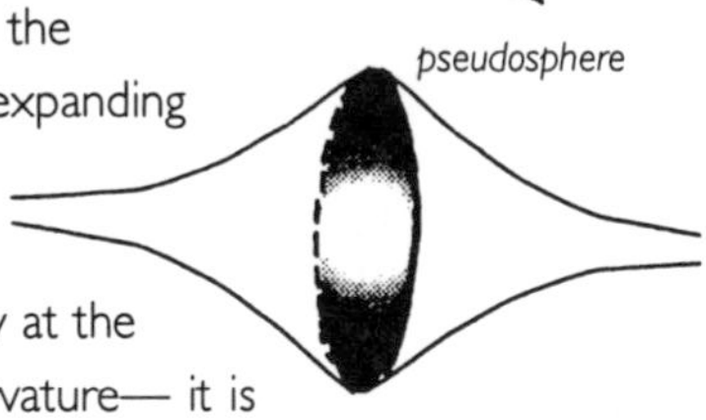

*pseudosphere*

Current astronomical findings indicate that the universe is expanding, but is it due to positive, negative or zero curvature? Will the expansion continue forever with celestial bodies moving away from one another, thus leading themselves to an oblivion of isolation into ever colder darker realms of the universe? Will the universe at some point shift gears and recede until it collapses on itself (the Big Crunch)? Or will it come to an almost standstill? Or perhaps the universe is an enormous yoyo in the hands of its creator? Data and theories are currently pointing to a negatively curved universe implying the universe is finite, but the constant influx of incredible information— such as exploding stars called type-Ia supernovas that hint at an antigravity force stretching the universe's fabric like taffy — is confusing what we think we know. What is certain is that we are on an incredible roller coaster ride of cosmological discoveries and theories, so hold on tightly.

# numbers help nail errors

It has been said that figures rule the world. Maybe. But I am sure that figures show us whether it is being ruled well or badly.

—Goethe

Have you ever wondered —how a bank clerk, entering a wrong digit in your account number, immediately knows something is amiss? —how transposed digits on a credit card entry come up *number invalid* rather than as someone else's account? —how a postal code scanner evaluates its reading of a letter's printed bar code?

All of the above involved a *check digit.* Check digits are attached to a code's number, and an algorithm utilizes it along with the code number to verify the validity of a product's code number. A product's code number — be it a bank account number, an airline ticket number, a social security number, a book bar code — is represented by a series of digits. Each item's digits fulfill specific criteria along with its check digit. Today, bar codes have become a way of life, appearing on everything from grocery items, to books, to tools, to toys. A bar code's numbers are transmitted via a scanner, magnetic strip or direct entry to a computer that processes the information and thereby identifies the object and checks the validity of the item's number. Utilizing this information, computer software helps businesses keep track of inventories and sales.

Bar codes are one of the most efficient, inexpensive and accurate ways to enter information into a computer. The scanner shoots pulses of light which land on the light and dark colored bars — light bars are converted to 0s and dark bars to 1s. The software takes the binary digits and performs an instantaneous data check utilizing a check digit. For example, for credit card number 4**6**7**3** 6**1**7 **4**8**9** 5**4**7, the first digit on the *right* end is the check digit, 7. It is determined by doubling the second number on the *left* end of the card number and every other number after it (don't include the check digit), yielding the numbers— 12; 6; 2; 8; 18; 8. Now add the digits of these numbers together: (1+2)+6+2+8+(1+8)+8=36. To this sum add the other digits in the card number, except the last digit on the right, which is the check digit we are trying to come up with. So we add, 4+7+6+7+8+5=37 **+36** =73. Now subtract this total from the next highest number ending in zero, namely 80. We get 80-73=7, which is this card's *check digit.* Using a method called mod 10 method, a credit card number is verified as follows—

1) Begin with the second digit from the right, double it and alternate digits, as we did before — **12, 6, 2, 8, 18, 8**

2) Add the digits of these numbers to the digits not doubled in the card's number. The sum of the doubled digits is (1+2) + 6 + 2 + 8 + (1+8) +8 =**36**. The sum of both sets of digits is:
4+ **(1+2)**+7 **+6**+6**+2**+7**+8**+8**+(1+8)** +5**+8**+7=80.
If this sum is divisible by 10 with no remainder, the card number is valid.

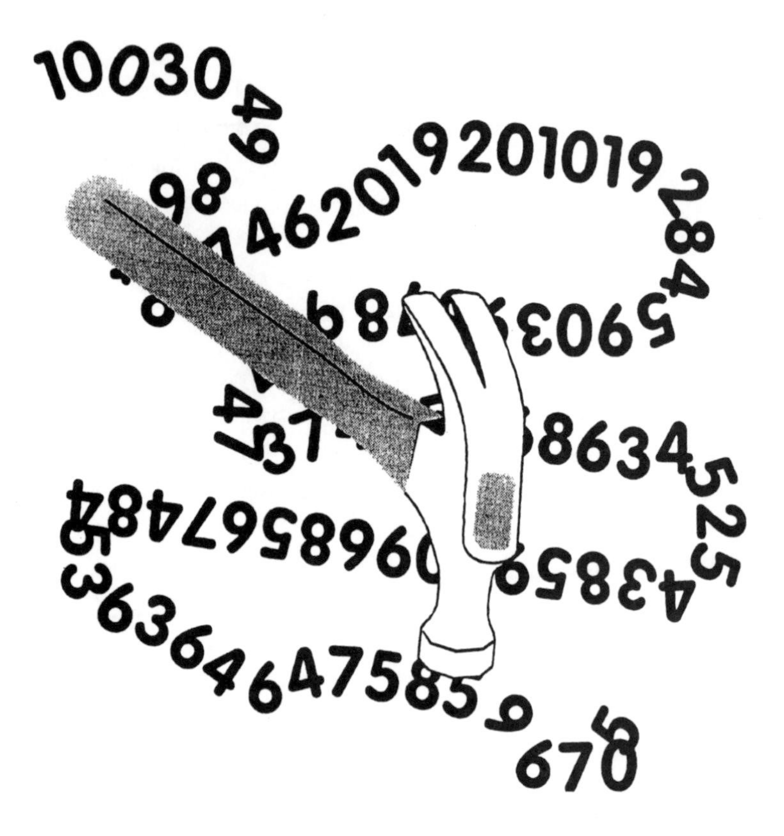

Zip codes also use check digits. A zip code has a check digit attached at the end of the bar code printed on envelopes.

> — The assigned digit is arrived at by doing the following algorithm. If your zip code is 94070-1755, just find the sum of these 9 digits and subtract that sum from the next highest number ending in zero. 9+4+0+7+0+1+7+5+5=38. So its *check digit* is 2, 40-38=2. Thus, the *Postnet* bar code for 94070-1755 has a check digit of 2 attached to its encoded form, 940701755**2**.

The check digits on bar codes decrease the chance of errors to 1 error occurring in every 10-million entries.

Check digits and number gymnastics play an important role in safeguarding against encoding errors that might occur, but the real enforcer behind catching errors is the field of mathematics called *informational theory* . For example, when a credit card is swiped so its magnetic strip is read, the information is transmitted in digital bits —zeros and ones— which may be altered by static, surges, damage or dirt to a strip, a disk, or to sound waves. Here is where mathematical error correcting codes, which detect and correct random errors, come to the rescue. Check digits are usually able to bring an error to our attention, but they are not capable of correcting the mistake. Some other check digit techniques are:

1) In an airline ticket number the last digit on the right is the check digit. Take the rest of the airline ticket number, and divide it by 7. If the remainder of this division matches the check digit, the ticket number is supposedly correct.

2) Most grocery store scanners are programmed to add every other digit of the product's code number starting from the digit on the left end, and then triple this result and add this to the remaining digits. If the answer does not end in zero, an error occurred.

In *information theory,* mathematicians have been developing methods for detecting and correcting messages which are being transmitted digitally via radio waves, telephone lines, computer disks. One method requires repetition of digital data, which is analogous to asking someone to repeat a comment to be certain it was heard correctly. For example, if 0110 is the data being sent, it could be sent as 00111100 with each of its digits repeated twice. Thus, if the transmitted message comes out 01111100, it is detected as an error. Then a computer program analyzes the digits and decides how to alter the digits so that the intended message is conveyed. Another method verifies a string of binary digits by attaching check digits at the end of the string's digits. In both cases computer software detects and corrects the error. Both methods must contend with the most economical use of time, but their accuracy rates are in the high 90s percentile. Linear codes are the error correcting methods that have been predominately used because they are easy to encode and decode. A code is a collection of codewords. A codeword is a string of 0s and 1s. A code is linear if it is the sum of any two codewords. On the other hand, nonlinear codes, which require less digital data, have been very difficult to encode and decode. In recent years, breakthroughs have been made combining binary digits (linear codes) with quaternary digits( 01,2,3) to write linear quaternary codes which end up being linked to nonlinear codes in relatively simple ways. Certain systems have also used ternary number systems. If simple ways can be refined to use nonlinear codes, the processing time for error detection and correction could be dramatically decreased. It seems almost ironic that such basic number systems— base two (binary), base three(ternary) and base four (quaternary) —are used to tackle the complicated problems of transmitting and receiving codes and error detection and correction. Whether you are buying a book, making a credit card purchase, or completing an ATM transaction, these simple 2, 3, and 4 digit number systems are at work for you.

The human mind has first to construct forms, independently, before we can find them in things.

—Albert Einstein

# mathematics & the art of Tony Robbin

What do quasicrystals, hyperspace, nonperiodic tilings, and 5 fold symmetry have in common? Tony Robbin — a contemporary artist whose work has brought to light in new ways the possibilities of higher dimensions and the symmetries of quasicrystals[1].

An artist works in a defined space. For painters, that space is the plane of a canvas. Over the centuries

[1] Until the 1980s, all crystals were supposedly periodic and did not have fivefold symmetry (an object has fivefold symmetry if its pattern can be matched up after it has been rotated 1/5 of the way around a circle, i.e. 72°). In 1982 chemist Daniel Shechtmann was able to form a super strong alloy from maganese and aluminum which had, the unheard of, fivefold symmetry. These discoveries led to a new class of crystals called quasicrystals, and a new general definition for crystals..

painters have transformed the canvas from 2-dimensional images to 3-dimensional ones, as witness the surreal flat images of Byzantine icons and the realistic scenes with perspective of Renaissance artists. All during the transformation of space on canvas, the mathematics of projective geometry has been present. Some artists, such as Cezanne, created the illusion of space beyond the canvas by distorting traditional 3-dimensional space with multiple points of perspective and by extending the periphery of his canvas by implying its existence beyond the borders of the canvas. (See *Still Life with Fruit Basket* (1888-1890)). Other artists, such as M.C. Escher, superimposed 2nd and 3rd dimensions in their work. In Escher's *Reptiles,* flat tessellated lizards become realistic 3-dimensional forms. After the introduction of non-Euclidean geometries and Georg Riemann's multiple dimensions and wormholes, the mathematical concepts of higher dimensions fired the imaginations of mathematicians, physicists and artists alike. At the turn of the 20th century, we have the cubist creating works with objects that exist in multiple dimensions and multiple time periods. Today, Tony Robbin's paintings redefine space once again. Instead of painting familiar objects in higher dimensions, the subject of many of his works is the 4th dimension. Through his eyes the viewer experiences hypercubes and other 4-D objects.

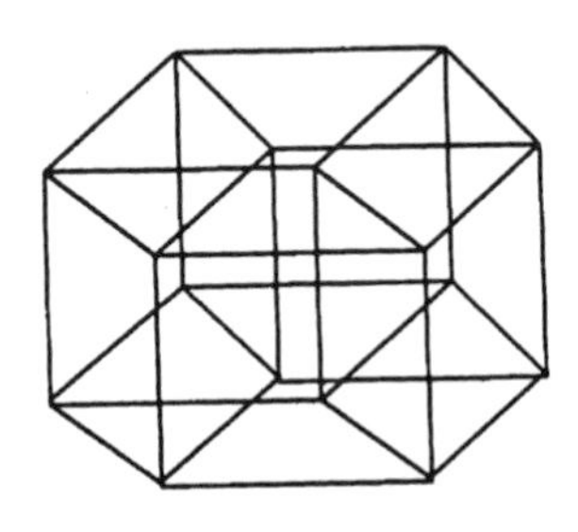

*One of the first published drawings of the hyperspace by William Stringham (1880). A 4-dimensional cube was first drawn by Swiss mathematician Ludwig Schläfli in the mid 19th*

For years, exploring and being able to visualize higher dimensions has not only been a quest for Robbin, it has been a passion which he has undertaken with the aid of mathematicians and computer scientists. He immersed himself in the mathematics of higher dimensions, and developed and used programs[2] to help him use the computer to

[2]Robbin developed the program FOURFIELD, which he initally used in his work, and is discussed in his book *Fourfield: Computers, Art & The 4th Dimension.* He now uses FORMIAN, developed for engineering in England.

***Getting glimpses of a hypercube:*** *Imagine a tree casting shadows of itself and its leaves. The wind comes up and the shadows begin to move and change in shape. In essence, this tree's shadows are 2-dimensional representations of the 3-dimensional tree. Each shadow is a projection of a part of the tree. Scientists visualize higher dimensions in a similar way, but since they do not physically have a hypercube to cast its shadows, they have written computer programs to look at cross sections of the hypercube as it is cut by planes. To visualize this process, imagine a sphere moving through a plane and the different impressions it leaves on the plane— a series of concentric circles with a center point. A cube would leave a variety of cross-sections — from a single point to a line segment, to triangles, to squares, to hexagons— depending on the possible angles of intersection. Putting together such information of these impressions gives insight into the physical form of higher dimensional objects.*

uncover the views and projections of the hypercube. His interest and persistence have created exciting works that give new visual meaning to the hypercube.

Robbin has used various techniques to convey the views of the hypercube. In *Fourfield,* welded steel rods and painted lines guide our view of the various planes and their rotations. Other paintings, such as *Simplex #9, Labofour,* and *Untitled #20* are done in acrylic on canvas with objects portrayed in various colors, shades, textures, and designs. Some of the objects in these works are layered on others and placed at angles that make you feel you are experiencing various dimensions simultaneously. One sometimes ends up wondering which dimension is actually being viewed.

Quasicrystals and nonperiodic tilings play important roles in Robbin's sculpture. In addition, shadows also have an important part in Robbin's metal with plastic sculptures.[3] The interconnection of forms lets the viewer realize that one dimension is not independent of another, but all exist at the same time, as in his *Quasicrystal Dome.* Imagine a quasicrystal dome made of rods emanating from the 12 faces of a

[3] Robbin created *Quasicrystal,* a larger than life size sculpture for the Technical University of Denmark. It is 50 feet long and 3 stories high

*Untitled #8 by Tony Robbin, 1980, 56"x80", acrylic on canvas, collection of the artist. Photograph courtesy of Tony Robbin.*

dodecahedron nodes connecting one another and occupying 3-dimensional space. Some of the faces of this sculpture are layered with colored plastic. The drama of the sculpture is enhanced by the experience of the amazing patterns of shadows cast on the floor —nonperiodic tessellations— and the transformation of shapes formed by the rods as the viewer walks under the sculpture. One moment one sees a series of triangles, while in another position the view becomes interlaced pentagonal stars. Being able to experience simultaneously two and three dimensional tessellations is truly amazing, and the implication of the 4th-dimension is an added experience. In addition to being artistic expressions, Robbin's sculptures provide a form of proof that quasicrystals actually form *periodically* in the 4th-dimension (see photograph below). It is

Model of *Quasicrystal Spaceframe,* 1989. Note that shadows form a Penrose tessellation. Photograph courtesy of Tony Robbin.

only when projected onto the floor (a plane) that their nonperiodic patterns and five-fold symmetry appear.[4]

When we think of space, we have been conditioned to think in terms of 3-dimensional space because that is the space in which we live and move about. An artist takes space and through his/her creativity creates it into the worlds of the mind's eye. Tony Robbin has combined art and mathematics and produced works of art which introduce us, in an aesthetically pleasing way, to the fascinating worlds of higher dimensions and other mathematical ideas.

---

[4]Crystallographers make projections of crystals and quasicrystals onto a plane by using X-rays. This is discussed in more detail in the chapter *It's Crystal Clear—quasicrystals & Penrose tiles.*

I know that two and two make four — and should be glad to prove it too if I could — though I must say if by any sort of process I could convert 2 & 2 into five it would give me much greater pleasure.

—Lord Byron

# Is this the last of **Fermat's last theorem?**

Using only **positive** whole numbers for x, y and z, how many solutions can you find for the equation, $x+y=z$?

How about $1+2=3$; $1+1=2$; $5+7=12$ ?

You are right! There are infinitely many solutions.

What about for the equation, $x^2+y^2=z^2$? As you surmised this looks like the Pythagorean Theorem equation. And again there are infinitely many solutions. Among these, 3, 4, 5 (since $3^2+4^2=5^2$); 6, 8,10; 9, 12, 15 ( in fact any multiple of 3, 4, 5) and any multiple of 5, 12, 13, etc. As before these are but a few of the infinite number of Pythagorean triplets.

Now, let's try the equation, $x^3+y^3=z^3$. Hmmm. I can't come up with any other than 0, 0, 0, which is excluded since 0 is not a positive number.

What about the equations $x^4+y^4=z^4$ ?

And for the equation $x^5+y^5=z^5$ ?

For $x^6+y^6=z^6$ ?...... For $x^n+y^n=z^n$? Any luck? Are there any higher exponents that work?

You have just been experimenting with Fermat's last theorem. In fact, no one has come up with *any positive whole numbers for x, y, and z when natural number n is greater than 2.* It all started in the 17th century, when French jurist Pierre de Fermat scribbled a statement in the margin of his copy of Diophantus' *Arithmetica* [1]. In addition to stating his conjecture, he tantalized mathematicians for centuries to follow by adding...*I have found a truly wonderful proof which this margin is too small to contain.*

*What came to pass?* Its proof has been churning in the minds of mathematicians and laypeople for over three and a half centuries. Prizes and awards have been offered for its proof.[2] The list of mathematicians attempting its proof and its ramifications reads like the who's who of mathematics. Among these we find—

Euler (1707-1783). Gauss (1777-1855) proved $x^3+y^3=z^3$ has no postive

---

[1] Fermat had a habit of making notes in book margins. In fact, the margins of his *Arithmetica* had many mathematical conjectures and ideas. His son published Fermat's copy of *Arithmetica* complete with his father's notes.

[2] The French Academy of Sciences offered a gold medal and 300 francs for the proof of Fermat's last theorem in 1815 and 1860. In 1909 Paul Wolfskehl bequeathed 100,000 marks for a published proof judged correct by the German Academy of Sciences. Within the first three years over a thousand proofs were reported to have been submitted. Even today some are still submitted, though the prize is now worth about $5000.

*Pierre de Fermat*

whole number solutions. Fermat had written a proof for $x^4+y^4=z^4$. The first proof for $x^5+y^5=z^5$ was given by Legendre (1752-1833). Dirichelt (1805-1859) proved it for $x^{14}+y^{14}=z^{14}$, while Lamé proved it for $x^7+y^7=z^7$. Sophie Germain made an important contribution toward the proof which dealt with the case $x^5+y^5=z^5$.[3] The next major inroad was made by Ernst Kummer in 1840 by proving Fermat's last theorem for another large group of numbers.[4] Among contemporary mathematicians in Fermat's last theorem hall of fame we find André Weil, Yutaka Taniyama,

---

[3] She proved if $x^5+y^5=z^5$, then either x, y, or z must be divisible by 5.

[4] He proved the theorem for all primes less than 100 except 37, 59, and 67. With the advent of the modern computer, computer techniques have been devised to tackle Fermat's last theorem. For example, in 1987, using computer technologies and methods, Samuel Wagstaff and Joanthan Tanner showed that Fermat's last theorem held for all exponents up through 150,000.

Taniyama, Gerhard Frey, Gerd Faltings, Goro Shimura, Kenneth Ribet, Barry Mazur, Andrew Wiles. They are some of the many mathematicians who had some success with Fermat's last theorem. Uncounted are those who tried without making any new discoveries or reaching any new results. But many important and involved theorems arose from attempts to prove Fermat's last theorem.

***Where does the proof of Fermat's last theorem stand today?*** It has been officially proven. In June 1993, Princeton professor Andrew Wiles shook the mathematics world and the general public when he climaxed his last lecture given at a conference at Cambridge with the announcement that he had proven a portion of the *Shimura-Taniyama conjecture*, which mathematicians felt held a key to proving Fermat's last theorem. His 200 page proof, entitled *Modular Elliptic Curves & Fermat's Last Theorem*, has since been scrutinized by Wiles himself and other mathematicians. Some significant gaps surfaced which as of June 1994 had not been resolved. But at the end of October 1994, Wiles surprised the mathematical community once more with two new manuscripts; one dealing with the problems that surfaced in his original proof and the other the mathematics justifying an important step of his proof. On this latter portion he collaborated with Cambridge mathematician Richard L. Taylor. Wiles' approach is an indirect proof — assuming Fermat's last theorem is false and seeing if a contradiction arises. Along this vein, a strange elliptic curve[5] surfaced, and, in addition, a contraditcion to the Taniyama-Shimura conjecture. Hence the reason why the proof of the Taniyama-Shimura conjecture is

---

[5] Elliptic curves (not to be confused with ellipses) are curves with equations of the form $y^2=x^3+ax^2+bx+c$, with constants a,b,c. When mathematician Yutaka Taniyama first suggested when elliptic equations with rational values for x and y produce the mathematical object called a modular form. Goro Shimura expanded on this idea, and hence the Taniyama-Shimura conjecture. Mathematicians initially felt these two object were so different that no connection between could exist. Now it looks as though a connection has surfaced.

important to Wiles' proof. The two papers were submitted to the *Annals of Mathematics* for publication, and have met with approval.

But this is not the last of the news on Fermat's Last theorem. In 1999 mathematicians Brian Conrad, Richard Taylor, Christophe Breuil and Fred Diamond worked on the Taniyama-Shimura conjecture for *all* elliptic equations, rather than just specific cases previously explored. Their work unveiled a connection between two mathematically different objects (elliptic curves and modular forms) thought to be totally unrelated. It seems that the proof of Fermat's Last theorem is a stepping stone for the proof of Taniyama-Shimura conjecture, and it now appears these four mathematicians may have proven the Taniyama-Shimura conjecture for all elliptic curves.

# mathematics & the game of life

*game theory*

Mathematics is a game played according to certain simple rules with meaningless marks on paper.

—David Hilbert

"Are you traveling together?" the ticket agent asked.

"Yes," you responded.

"So how's row 7— aisle and center or center and window?" the agent asked.

"No, we'd like row 7, aisle and window," you replied. The agent gave you a strange look, but decided not to question your motives and accommodated your request.

What were your motives? You were banking on having extra room during the flight by assuming a single traveler will choose either an aisle or a window before selecting a center seat. You were participating in *game theory*.[1] Whether you are buying a car, selling your home, playing poker— the principles of game theory are at work. Whether two countries are negotiating a trade agreement, settling a border dispute or in the midst of an arms race —game theory strategies come into play.

The famous *prisoners' dilemma* problem was first proposed by Albert Tucker in 1950. Two people are arrested, and placed in separate rooms for questioning. Each is immediately confronted with the following options: (a) If you and your partner remain silent, there's enough evidence to give you each 2 years in prison. (b) If you confess and your partner doesn't, you get off completely and your partner gets 4 years in prison. (c) If you both confess, you both get 3 years in prison.

Can you take the chance and remain silent? That would mean either 4 or 2 years in prison. Should you confess — that gives either 3 years in prison or getting off completely. This situation is a one-time choice for each suspect to make. If you have no idea how your partner will act, your best option is to confess. But when you have an opportunity to "play" the "game" repeatedly, your actions influence your opponent's actions. Messages are communicated about how you will respond. For example, companies X and Y are just beginning to do business. X supplies products to Y. X fulfilled Y's order and billed Y for their order. The terms were payable upon receipt. A month passes, and the invoice is not paid. Does X rebill Y with added interest or first call Y's bookkeeping? X calls and learns Y supposedly never received the bill,

---

[1] *Game theory* was formally introduced by John von Neumann and Oskar Morgenstern in 1944 in their book *The Theory of Games and Economic Behavior.* Since then, mathematicians have been analyzing and mathematizing games be they board games, card games, or games of life.

A computer game called *Life*, invented by John Conway in 1970, illustrates elements of *game theory*. The game begins by selecting some initial pixels on the computer screen. Next, the rules Conway established with regard to empty and occupied pixels are applied, and the universe of the computer screen begins to fill up with interesting patterns.

and in fact Y wants to reorder. X says the order will be sent as soon as Y's bill is paid. With the second order, Y again does not pay upon receipt. Now X does not give a courtesy call or grace period, but rebills Y with finance charges included. This is your typical *tit for tat* situation. The best game strategy here is to always first cooperate to see how your opponent acts. Then your next action is based upon whatever your opponent's previous action was. This situation initially involves cooperation, then provocation, followed by a suitable response, which hopefully will be followed by cooperation again.

Even the game of *chicken* lends itself to mathematizing. Here drivers race toward one another. The first driver to swerve is the loser. Possible outcomes: both have an accident because neither driver swerves off and neither wins —both drivers swerve simultaneously, neither wins nor loses face (it is a tie) —one driver swerves before the other, who is the winner. To mathematize this game, number values are assigned to the possible outcomes — 0 points for a collision, 2 for a win and 1 for a swerve, and these can be arranged in a table.

*Game theory* pops up in all sorts of areas — in the environment with lobbyists vying for legislation for clean air and against polluters— with medical insurance companies deciding who should receive what treatment. Have you ever cut into line? Were you successful, or did someone call you on it? Or have you asked a person in front of you if you could go ahead with your one grocery item? That's *game theory* at work. *Game theory* can sometimes bespeak cooperation, and has shown that mutual cooperation in the long term fosters the most favorable outcomes.

Choosing to contribute to public radio or television is your choice. If you decide to tune in or watch your favorite show, that's also your choice. If you then choose not to contribute, you are banking on the other people paying for you. If on the other hand, the station goes belly up you can't complain because you made the wrong choice in your game of life. Time and again the elements of *game theory* replay themselves where conflicting forces are at odds — in labor strikes, political disputes, terrorist/hostage showdowns, nature. Mathematically analyzing the possible outcomes/payoffs and possible choices, action, reactions may help in the decision process. Unfortunately, in the "game of life" it is not possible to identify and assign a number value to all factors that impact the outcomes. Even using probability, human behavior may be unpredictable; and where personalities and egos come into the picture the result may lead to chaos. In the final analysis, the game isn't over until it's over!

The essence of mathematics is its freedom.

—Georg Cantor

# mathematics & the architecture of SFMOMA

*San Francisco Musuem of Modern Art*

Today's architects are creating designs impossible in the past because certain mathematics, materials and tools did not exist.

Be it the Empire State building or the Guggenheim Museum in Bilbao, people are awed by architecture. In 1996, it was the San Francisco Museum of Modern Art (SFMOMA). From the date of its opening, the SFMOMA has been a focus of attention. Museum goers from all over line up daily to get an interior glimpse of this architectural work. Standing out from the neighboring highrises, the SFMOMA's exterior exquisitely balances geometric shapes, symmetry and unusual facades, while its interior offers special feasts for the eyes. According to the designer, Swiss architect Mario Botta, "the real challenge was to discover that perfect balance where the architecture and art enrich one another." To achieve his creation Botta employed the classic notion of line symmetry. A vertical line divides the frontal view of the building into two identical halves. It is striking to see this symmetry used with such a diverse combination of geometric shapes. In the front view perspective, the structure presents the 2-dimensional shapes of rectangles, squares, circles and ellipses, while the sides and back add triangles and semi-circles. This unusual combination of flat shapes — circles rectangles and triangles — adds excitement to the structure as the viewer anticipates how these shapes will play themselves out in their 3-dimensional form in both the exterior and interior. When approaching the museum, one is struck by the fortress-like quality of its exterior— bold massive rectangular solids symmetrically stacked and constructed from earthtone bricks, and crowned by a spectacular truncated cylinder, set off by contrasting black and white granite strips. The noticeable ab-

*Front view of SFMOMA. Courtesy of the SFMOMA*

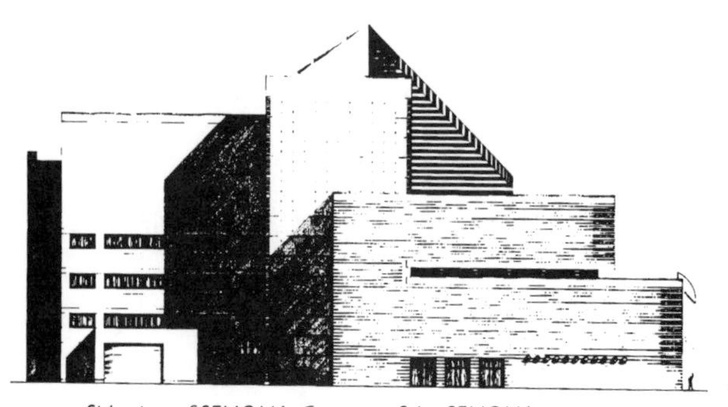

*Side view of SFMOMA. Courtesy of the SFMOMA*

*The San Francisco Museum of Modern Art. Photograph by R. Barnes. Courtesy of SFMOMA, © 1994 SFMOMA.*

sence of windows carries out this exterior fortress-like facade. Upon entering, one is immediately impressed by the infusion of natural light streaming in via the truncated cylindrical skylight. At night the role of this skylight reverses, and it becomes a beacon of light for the city. The angle at which this cylinder was truncated maximizes the sunlight entering the interior's central plaza. As Botta points out "The atrium, the true heart of the building, is the center of spatial gravity for the entire Museum. Within its interior, the organization and spatial relationships of all those parts that surround and define should be perceived. ...It is an architectonically drawn space, with the light from above serving as a type of cornice for ...the Museum visitor. The galleries, on upper levels, offer subdued and calm architecture, which,...devote their space to...the work of art."[1] Entering the museum, one finds the five floors skillfully concealed

[1] SFMOMA press release: *Mario Botta: The Architect & His Philosophy.*

*Interior stairwell of San Francisco Museum of Modern Art. Photograph by R. Barnes. Courtesy of SFMOMA, © 1994 SFMOMA.*

by expansive ceilings and staggered levels. As one begins to explore the interior, new floors effortlessly emerge and invite the visitor to enter via the bold black granite stairwell. The staircase of the central plaza is both a work of art and an optical illusion. The undersides of the stairwell play with the contrasting black and gray granite strips of the walls creating an oscillation type optical illusion, making the underside flip back and forth in one's mind, in much the same way objects in Escher's drawing *Concave/Convex* trick the mind. The museum is designed with only a few lateral windows, thereby allowing maximum wall space for displays. Yet, Botta's architectural design stills allows natural light to abound on all floors in most of its galleries because of the ingenious way adjustable skylights with light diffusers and curved translucent paneling were specifically engineered.[2] To achieve this optimum lighting, Botta studied the light by working with a scale model of the museum at the site and observing how the light hit it during different times of the day. Although the museum was designed to house modern art, it is itself a work of art with a treasure trove of mathematical objects and ideas at work.

---

[2]Flack & Kurtz Consulting Engineers developed these skylights over a three year period.

# music, matter & mathematics

Every new body of discovery is mathematical in form, because there is no other guidance we can have.

—Charles Darwin

Have you ever been in a room and sensed something affecting you? Maybe it was the buzzing of a fountain's motor, the shrill sound of a passing siren, an enchanting melody, or various sounds, noises, vibrations — a leaf blower, a buzz saw, a bird's song. Some sounds are pleasing, others aggravating; some make us feel uncomfortable.

References to the power of music and sound go back centuries. In *The Odyssey,* we learn of the songs of the Sirens, and of the lives they claimed. Also in *The Odyssey,* Circe warned Odysseus of the dangers of the Sirens' voices when she said: "...whosoever draws too close… the high thrilling song of the Sirens will transfix him, lolling there in their meadow, round them heaps of corpses…".[1] Aristotle stated: "The motion of bodies…produce a noise. …The sun and moon… and all the stars … moving with so rapid motion, how should they not produce a sound immensely great? Starting from this argument, and from the observation that their speed, as measured by their distances, are in the same ratio as musical concordance , … the sound given the circular movement of the stars is a harmony."[2] The ancient Pythagoreans discovered there was an innate correlation between music and mathematics as specifically displayed in the number ratios of the octave. They felt that numbers linked with music were the means by which the universe was ordered, contending the movement of celestial bodies produced musical sounds which were determined by the bodies' speeds and distances from the Earth. Although the idea of the *music of the spheres* originated with the Pythagoreans, it influenced such scientists as Johann Kepler, who linked the velocities of the planets in their elliptical orbits with musical harmony, and wrote music for each of the then known planets.

There are many discoveries being made today about the influence of music upon many aspects of our lives. One study reveals that formal music instruction can have a positive affect on a child's spatial intelligence (Science News August 27, 1994, *Tuning up young brains*). A University of California at Irving study showed that students listening to 10 minutes of Mozart's *Sonata for Two Pianos in D Major* (K. 448) raised their IQ scores on spatial temporal reasoning (skills related to mathematics) tests. How does one explain that? It is believed that Mozart strengthens neural connections that underlie mathematical thinking. In addition, some researchers found this sonata also reduced seizures in epileptics and improved spatial temporal reasoning in Alzheimer patients. We know

---

[1] *The Odyssey, by Homer* translated by Robert Fagels. p. 272

[2] *The Works of Aristotle*, volume 2, translated by J.L. Stocks, Oxford University Press, 1930.

our bodies react to sounds, songs, music and vibrations in various physiological ways. Loud unpleasant sounds can cause our blood pressure to rise, constrict blood vessels, speed up our heart and breathing rates. The pounding of a base drum in a passing band of a parade can affect one's heart rhythm. Sounds can even alter the levels of fats and magnesium in our bloodstreams. The volume of noise is not necessarily the disturbing factor. Even silence and soft dissonant sounds can have negative effects.

**strings vibrating in the universe... determining matter and energy... compacted in 10 and 26 dimesnions**

Music is all around us, and it is mathematics that helps bring its pervasiveness to the surface. It explains how sound travels from its source in 3-dimensional space, and utilizes its characteristics of pitch, loudness, and quality to create curves and equations to describe sounds (sinus equations[3]). Mathematicians have expanded the idea of the sinus curves to wavelets which are now being used to describe any vibration or picture. Scientists use these equations and visual descriptions to chart earthquakes, evaluate heart rhythm, and even describe fingerprints.

---

[3]In these equations, pitch is related to frequency, loudness to amplitude, and quality to the shape of the periodic curve

Today, physicists have developed incredible ideas about the influence of music in the universe. But now it is no longer the music of the spheres, but the music of the strings — superstrings that is. These infinitesimal strings may hold the secrets to the essence of matter and energy by the way in which they vibrate. String Theory[4] contends that the vibrations of these superstrings are the building blocks of the universe, distinguishing one form of matter and energy from another. Where do these strings reside? Everywhere in the universe, in the depths of all matter and energy — compacted[5] in multiple dimensions of geometric frameworks whose nature still remains a mystery, but whose mathematics beautifully fits together with the ideas of physics. There are many versions of the String Theory. In one, the strings are called hererotic, and are described as closed strings vibrating clockwise or counterclockwise. The clockwise reside in a 10-dimensional space, while the counterclockwise are in a 26-dimensional space where 16 dimensions are compacted. String theories deal with 10 and 26 dimensional spaces and are the only quantum theories that require a fixed number for the dimensions of space-time. Why 10 and 26? Physicists don't have a clue.[6] It is just that the mathematics of these strings and their vibrations work perfectly and encompass Einstein's equations and the theory of gravity — it reconciles Einstein's theory and quantum theory without mathematical hitches. So why isn't one of these string theories adopted? There is no way, as of yet, to prove them. Many theories are constantly evolving about the universe, but the only legitimate way to choose one is to have proof. And in science proof does not mean that all the equations work out and

---

[4]There are numerous versions of string theory, but they owe their heritage to the work of John Schwarz of the California Institute of Technology and Michael Green of Queen Mary's College in London. These two men proved that all self-consistency conditions can be met by string theory.

[5]To get a feeling for a compacted world imagine a world of zero-dimension. Suppose the 0-D world actually resides in a 3-D world that is compressed into what appears to be a single point. The 3-D sphere's radius is so small, it cannot be measured, so it appears to be zero-dimensional.

[6] These numbers first appeared in the modular functions work of Indian mathematician Srinivasa Ramanujan, and the *Ramanujan function* repeatedly appears in string theory calculations.

fit together, nor does it only mean a mathematical proof — it means concrete observable evidence.

What are some other *mathematics-music connections*?

—The DNA double helix has been put to music by using the four bases comprising DNA. A musical link becomes apparent, when one considers the recurring sequences as recurring melodies of a song. These "musical" patttems have been explored by Dr. Susumo Ohno of the Department of Theoretical Biology at the Beckman Research Institute of the City of Hope at Duarte, Califomia.

—Mathematical curves play important roles in the shapes of many musical instruments, including string, percussion and wind instruments.

—Computer modeling and digitizing music are big fields. Coupled with acoustics, mathematics is being used to determine the shapes for acoustical architecture.

—Mathematicians are putting fractals to music. Some recursive numerical sequences, such as

0,1,1,2,1,2,2,3,1,2,2,3,2,3,3,4...

the sequence that counts the numbers of ones in consecutive binary numbers, are put to music. Musical pieces by composers are being studied by mathematicians for inherent fractal qualities. Physicist and modem composer Gyorgy Ligeti has deliberately used fractals in some of his compositions, such as in his *Etudes for Piano*. Other fractal properties are being studied in music.

Where there is music there is mathematics. In the words of James Joseph Sylvester, *May not music be described as the mathematics of sense, and mathematics the music of reason?"*

# nano-technology

> There is no smallest among the small and no largest among the large; but always something still smaller and something still larger.
>
> —Anaxagoras

Mathematics has always had objects that have been minute. The points of Euclidean geometry have no dimension. Technically they are invisible and only indicate the location of an object. In fact, a line segment is composed of infinitely many points. Between any two another always can be found. The world of calculus relies on the infinitely small to solve many of its problems. Fractal worlds have been the host of the ever duplicating self-similar fractals that have no smallest shape. The world of numbers has never been at loss for small quantities, since there is no smallest and no largest. Thus, mathematics is a perfect media for

*A schematic illustration showing a series of hydrogen deposition tools (above) placing hydrogen atoms on a series of cylindrical work pieces (below). © Copyright 1993 K. Eric Drexler. All rights reserved.*

quantifying and describing the objects of the nanoworlds[1]. When it comes to nanotechnology, the saying the world is getting smaller and smaller is no longer an empty phrase. Nanotechnology and all its possible ramifications boggle the mind. Could we have envisioned fifty years ago the products that would be generated from digital technology? Who would have guessed the impact of computers on every aspect of our everyday lives from medicine to banking to consumption to criminology to entertainment to communication to... Today, the average person can own and operate cell phones, scanners, answering and fax machines, personal computers, audio CDs, camcorders, microwaves, smart cards.... Will nanotechnology bring similar revolutionary changes?

Years ago, who would have believed a camera fed down your throat to photograph your stomach and intestines or a computer chip so small that microcomputers could replace the cumbersome mainframe

---

[1] To grasp the size of things in nanoworlds— a nanosecond is one-billionth of a second. An electric impulse takes a nanosecond to travel 8 inches. Light travels one foot in a nanosecond. A nanometer is one-billionth of a meter.

that microcomputers could replace the cumbersome mainframe computers of the 1960s? Now imagine a robot so small it is invisible to the naked eye, or a computer the size of a bacterium. Too far fetched? Along with modern computers, the prefix nano- has entered the 20th century with a host of new terminology. This partial list of nanoterminology— •nanotechnology • nanomachines • nanosubmarines • nanoscopes • nanotubes • nanocomponents • nanocircuitry • nanoassemblers • nanostructures • nanoengineering —emphasizes that the world of the ultra-minute is entering the mainstream, whether one believes it or not.

***What exactly is nanoworld[2] ?*** It is a world that functions on a scale the size of atoms and molecules. Nanotechnology encompasses molecular machines and gadgets that manipulate and form matter— matter of all types (animate and inanimate) — atom by atom and molecule by molecule. Richard Feynman spoke of such a world when he envisioned machines designed to make ever smaller machines again and again. However, much of the work in today's nanotechnology is following a route envisioned by Eric Drexler. Instead of starting from the top and working down, Drexler imagined working directly with atoms and molecules and building ultra small machines. Nature forms the various types of matter on the atomic level by using such things as DNA, proteins[3], lipids, peptides,etc.. Drexler was creative enough to believe the same would be possible in a nanoworld by manipulating atoms and molecules into a myriad of forms using various nanotools.

**What are the sizes?**

- ***micrometer*** **is a millionth of a meter.**
- ***nanometer*** **is a billionth of a meter.**
- ***picometer*** **is a trillionth of a meter.**

---

[2] Technically nano- means one-billionth the size, so a nanometer is 1/1,000,000,000 of a meter.

[3] For example, proteins are one of nature's nanomachines that can manipulate atoms in a substance and replicate themselves using the codes and programs of DNA

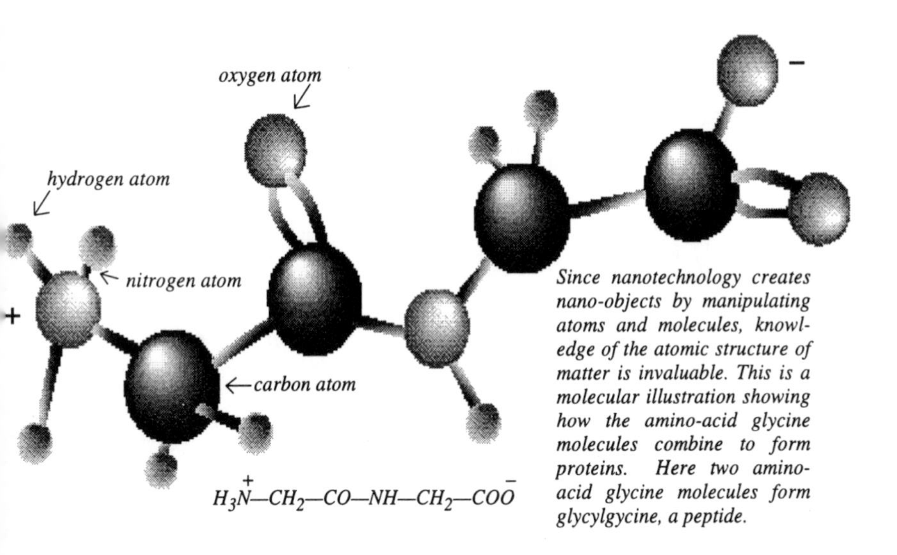


*Since nanotechnology creates nano-objects by manipulating atoms and molecules, knowledge of the atomic structure of matter is invaluable. This is a molecular illustration showing how the amino-acid glycine molecules combine to form proteins. Here two amino-acid glycine molecules form glycylgycine, a peptide.*

Even though the idea of nanotechnology has been brewing for only a few decades, nanoscientists, nanoengineers, and nanoreseachers have made significant progress and have actually constructed things of an atomic size scale.

***How do scientists go about building things they cannot see?*** Although atoms, molecules and other nanocomponents are invisible to the naked eye, new technology brings them within grasp. Using knowledge of molecular structures, coupled with advanced technological tools, molecules and atoms can now be manipulated. Armed with a variety of scanning probe microscopes[4], researchers can push, probe, and relocate atoms and molecules along a "flat" surface at room temperature. Of course, it was not as easy as it sounds, because techniques had to be developed to prevent molecules from jumping around when agitated by changes in temperature. In addition, virtual reality techniques have enhanced visual work so that scientists can experience an illusionary

---

[4] These include STM (scanning tunneling microscopes), SPM (scanning probe microscopes), AFM (atomic force microscopes), MRFM (magnetic resonance force microscopy).

3-dimensional view of the molecules in space. Work is being done to devise means to actually monitor work on a 3-dimensional level. Nano-research is taking place simultaneously in biology, chemistry, and physics, and mathematics is indispensable in this research.

***What has been made thus far in nanotechnology?***

Five years ago nanomachines and nanotools existed only as computer simulation models. Today —

- Researchers at Delt University of Technology in the Netherlands have made a tweezer-like tool capable of "picking up" clusters of nanometer-sized particles by trapping them between two electrodes. Researchers at Arizona State University have also developed nanotweezers by adapting the use of a STM. By arming them with a voltage pulse, atoms are coaxed and pulled up to its tip. Various measurements taken during this procedure are used to identify the atoms.
- Nanotubes and nanobearings have been made with carbon atoms. Scientists have found that some conduct electricity while others block its passage. Molecular wheels and propeller shaped molecules have been fabricated, which can rotate rapidly at room temperature on a circular bearing-like structure.
- Ring shaped protein fragments have been formed into peptide nanotubes. These can act as channels for ions and molecules to pass through, which may, in the future, be used to deliver antibiotics to specific cells.
- In biology, nanobacteria (ranging in size from 50 to 500 nanometers) have been discovered in kidney stones. These findings are expected to be useful in studying disorders such as atherosclerosis, cancer, arthritis, and unexplained calcium deposits in the human body. The normal bacteria's size is about 1 micromenter (or 100 nanometers).
- Nanotubes have been linked into chains of ring-shaped glucose molecules called cyclodextrin. Tubes are essential to living

organisms. Microtubes in cells function as conveyor belts that move different compounds around. In biochemistry they can be used in drug delivery systems, as a means to separate ions and molecules, or even as catalysts. Commercially, because of their size, they can greatly improve moisture retention in cosmetics.

- An efficient electrochemical cultivation method for bacteria has been developed.
- Various scanning probe microscopes have been adapted to work under a variety of conditions and temperatures. For example, an STM has been used as an atomic knife. AFM tips have been used to sense softening of cell walls which indicate a virus penetration. MRFM work is progressing to develop 3-D atomic imaging which will enable scientists to visually locate individual atoms.
- In order to demonstrate the technology of nanofabrication, researchers at Cornell University made a microguitar, whose strings are 100 atoms wide. Even a nanoabacus has been fabricated using buckyballs as its beads.
- 18th century French physicist Charles Augustin de Columb's electrometer has been scaled down to the size of a few micrometers.

*What's to come?*

The possibilities that nanotechnology offer are amazing and incredible. Among these are:

- Nanomachines whose shapes can change, i.e. they can undergo morphing to suit the task at hand by simply changing the software of the units. These will be like fractal shape-shifting robots able to adapt to the task at hand.
- Biological or synthetic nanomachines, which are self-replicating and self-assembling.
- Nanotech researchers have envisioned an Earth completely revolutionized by the science of nanotechnology. An Earth on

> which nanocomputers guide the assembly lines of trillions of nanounits. Nanomachines would be able to self-assemble and self-replicate. Fractal shifting robots would morph their shapes and uses to tackle any task at hand by simply changing their software. Tasks would include— cleaning toxic wastes (biological units have already been used in this area) and pollution, delivering drugs directly to diseased cells, repairing injured cells— nanogardening units would clear a yard of weeds and prepare enriched soil— nanohousekeeping units would dust and clean — foods would be assembled from their atomic state, as in *Star Trek's* replicator — waste and garbage would be recycled by nanounits returning its form to atoms and molecules, which would be used later to make other products. The scenarios go on and on.

An Earth completely governed by nanoscience would have no famine, no disease, no shortages. No problems? Maybe. Cannot these nanocomputers have nanoglitches, which may be thought of as nanomutations? Social and mental adjustment problems would be great. Today, we see many people who have difficulty accepting and relating to something they can actually see and hold—the computer. Yet, with modern medical diagnostic techniques using cameras so small they can be fed though a tube down the throat or in an artery through the groin, plus the advent of genetic engineering, minute invisible worlds are no longer foreign to people.

Is nanoworld as small as it gets or can get? No. There is talk of picotechnology. But perhaps the world of superstrings will scale down nanounits and picounits even further by using strings as building blocks of matter— strings compressed in multiple dimensions whose essence is not defined by atoms but by their vibrations. We can be certain that nanoworld will definitely be one of the steps to the world of the future.

# soft computing

*an emerging new way to process computer information*

Man has within a single generataion found himself sharing the world with a strange new species: the computer...

—Marvin Minsky

Soft computing is probably one of the most difficult types of computing to develop. Definitely not for softies, it involves the merging and utilization of high powered mathematics. Soft computing cannot rely simply on traditional "yes and no" logic. Here a statement is no longer just considered true or false, but its degree of truth or falsehood is taken into account. Here a situation may not have just two options — being or not being. The uncertainty of a situation can also enter the picture. Soft computing will involve cognitive reasoning with

degrees of vagueness and uncertainty built into it. *But what exactly is soft computing?* It is so new that it has not been clearly defined — but rather described in terms of *not* having the characteristics of hard computing. True or false logic, precision and speed are the primary characteristics of hard computing. Fuzzy logic, neural networks, probabilistic reasoning, ease of use and creativity describe soft computing. Soft computing will be able to deal with elements of imprecision, uncertainty and partial truths that until now hard computing has not.

*How will it do this?* As it evolves, soft computing will rely on mathematics from such fields as probability, fuzzy logic, chaos theory, complexity, neuroscience, and fractals. At a glance all these areas seem very different, but mathematicians and scientists are discovering cross uses of these expanding fields.

***What are these fields, and what do they deal with?*** In a nutshell—

***fuzzy logic***— considers the degree of truth. Unlike traditional logic, which assigns a true or false label to a statement, fuzzy logic can assign a value in between truth or falsehood. It provides room for gray areas.

***chaos theory***— studies the order that surfaces in chaotic phenomena. Chaos may be present in the simplest or most complicated phenomena from a dripping faucet to weather systems. Future outcome of a particular phenomenon is unpredictable because minute (perhaps imperceptible) variations in initial conditions can cause monumental results. (This is often called the butterfly effect.)

***non-linear dynamic system***— Linear systems are predictable. The same imputed data always produces the same results. With non-linear systems, specific data can produce different results. Ever changing conditions are influential in non-linear dynamic systems.

***complexity or a complex system***— operates in the realm of *non-linear mathematics.* Here the same set of circumstances does not always produce the same outcome or solution. A complex system possesses structure and order, while trying to maintain its delicate balance between chaos and order. If a complex system suddenly loses its balance, it can regain its equilibrium by its spontaneously self-organizing dynamics (i.e .by constantly changing and adapting itself to changing factors or circumstances).

***neuroscience*** — is a science that deals with biophysiological information and neuron networks.

***fractal***— is an ever evolving/growing object created by starting with a basic object or pattern to which a rule is applied over and over replicating the pattern in smaller versions ad infinitum, or it may infinitely replicate a changing pattern. The iteration of a process, a rule, or an equation plays a major role in producing a fractal. As a result, fractal geometry, a non-Euclidean geometry, can simulate almost any shape you can imagine.

***genetic algorithm*** — are algorithms by which computers learn and adapt information following Darwin's rule of natural selection — survival of the fittest. Fittest is defined by the desired goal of a particular problem; for example, the problem may be to optimize the fuel efficiency of an engine. A genetic algorithm is designed along lines analogous to a genetic code acting as living organisms. Some parts of the code merge and new forms are created. Some parts mutate. They compete. Some become extinct and the part producing the desired goal, in this case best fuel efficiency, survives as the solution. In other words, software is designed so that as it evolves it improves itself.[1]

"...soft computing is not a melange of fuzzy logic, neural networks theory, and probabilistic reasoning. Rather, it is a partnership in which each partner contributes a distinct methodology for addressing problems ..." says Professor Lotfi A. Zadeh creator of fuzzy sets.[2] Since hard computing is not capable of controlling or dealing with complex systems, fuzzy logic and fuzzy expressions will be needed to describe the systems found in nature and human behavior. In order to create soft machines, computers will need to integrate fuzzy logic, neuroscience, and probabilistic reasoning. As Takeshi Yamakawa says, "Computers are no

---

[1] Genetic algorithms were invented by John Holland of the University of Michigan in the 1960s. Today these algorithms are being commercially used in such areas as production scheduling and product designing. Their multiple uses are just evolving.

[2] An Internet message posted March 3, 1995 by Professor Lotfi Zadeh on which Berkeley Initiative in Soft Computing (BISC) mentioned the celebration of its third birthday on March 13, 1994.

good unless they are grounded in psychology and philosophy. If we think of computers as no more than an extension of logic circuits, Boolean variables, transistors technology, circuit integration, and microprocessing, then we are never going to create a computer that thinks creatively ... to create computers ...capable of recognizing human emotional patterns... we have a long way to go."[3] Professor Zadeh points out that "In many ways, soft computing represents a significant paradigm shift in the aims of computing — a shift which reflects the fact that the human mind, unlike present day computers, possesses a remarkable ability to store and process information which is pervasively imprecise, uncertain and lacking categoricity."[4]

---

[3]Takeshi Yamakawa, Dean of the Kyushu Institute of Technology in Japan, is considered a "fuzzy" scientist. See *Look Japan* December 1994.

[4]An Internet message posted March 3, 1995 by Professor Lotfi Zadeh on which Berkeley Initiative in Soft Computing (BISC) mentioned the celebration of its third birthday on March 13, 1994.

# it's crystal clear, or is it?

*quasicrystals & Penrose tiles*

The further a mathematical thing is developed, the more harmoniously and uniformly does its construction proceed, and unsuspected relations are discovered between hitherto separated branches of the universe.
—David Hilbert

Ever go to do a job and in the process found the job was not as simple as you thought? It seemed that one thing always led to another, and that the other thing was oftentimes something new or totally unexpected. Such was the case with nonperiodic tilings and crystals; two seemingly unrelated and unconnected areas.

As far back as ancient times tiling (aka tessellating) was used to adorn walls and floors. Over the centuries, hoping to facilitate their works, artists and craftspeople sought to understand the characteristics of tessellations, and mathematicians sought to explain the mathematical concepts behind them. Here we find the concepts of symmetry, periodic and nonperiodic tiling.

---

*What do these terms mean?*

Two plane objects are *symmetrical* to one another, if by either reflecting one about a line, rotating it about a point, translating(moving) it in another direction, or gliding it, they match up perfectly. Equiangular triangles, squares, and regular hexagons are the only regular polygons that can be tiled independently because their edges come together so no gap is left (totaling 360 ° at the junction of the vertices). Six equilateral triangles do this, so do 4 squares, and, likewise, 3 hexagons. All other regular polygons leave gaps.

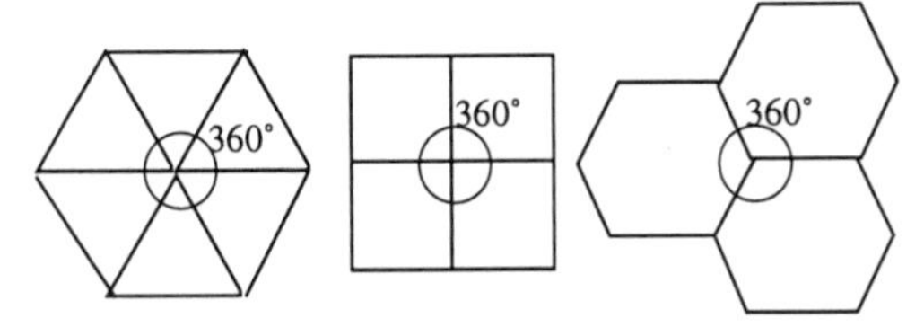


*Semiregular tiling* uses two objects to create a tessellation or tiling. The diagram, on the right, illustrates this using an octagon and a square. In semiregular tiling, mathematics has shown that no polygon can have more than 12 sides. In addition, semiregular tiling with polygons of 5, 7, 9, 10, 11 sides is not possible. In a *periodic tiling* the design repeats itself continually when the eye moves vertically or horizontally.

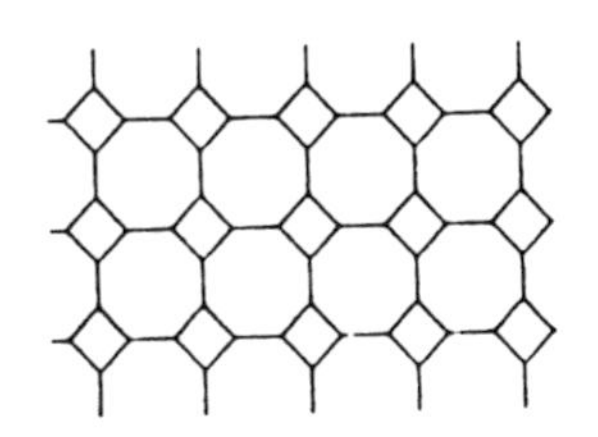

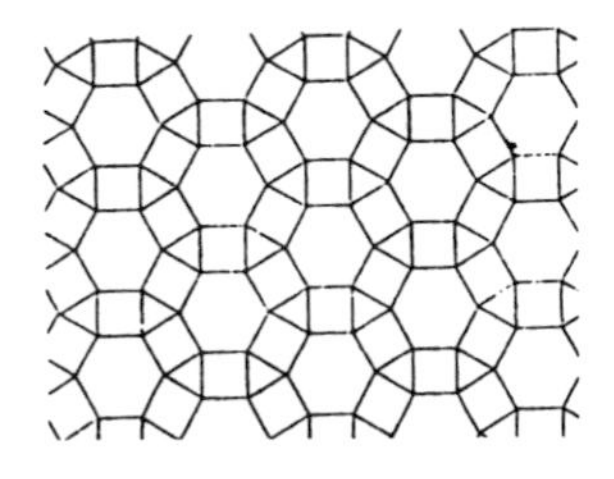

Mathematicians believed that if a nonperiodic tiling could be made with certain shapes, then a periodic one could also be made with those shapes.

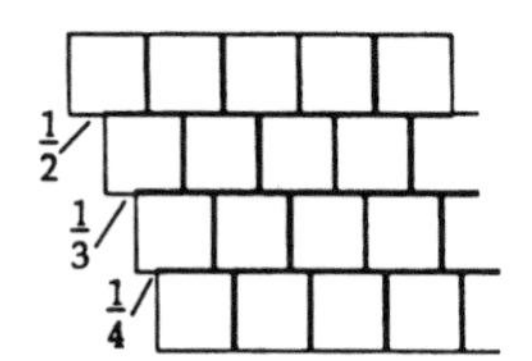


*A non-periodic tiling made using square tiles placed in staggered rows.*

In 1964 a set of 20,000 different tile shapes was discovered that only allowed nonperiodic tiling. This discovery led to other discoveries of sets with fewer distinct shapes that only produced nonperiodic tilings. Then, in 1974 Roger Penrose discovered a set of only two tiles which produce nonperiodic tiles, now called Penrose tiles. Penrose tiles possess a type of symmetry called *fivefold (rotational) symmetry* (and also tenfold symmetry). *Fivefold symmetry* means the tiling pattern can be matched up to another on the plane after it is rotated 1/5 of the way around. Such symmetry is found in the pattern on a sand dollar or on starfish.

*The two Penrose tiles.*

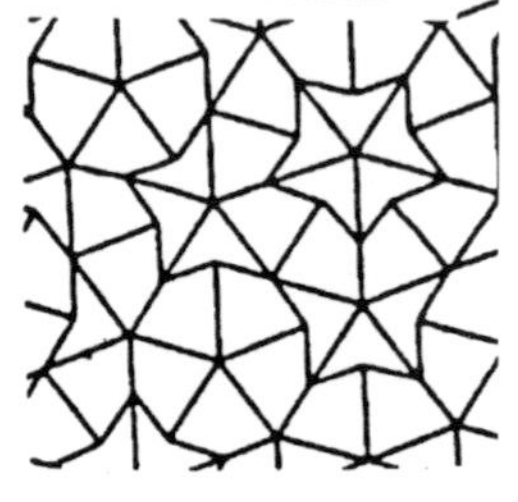

---

In the 3rd-dimension, tessellations are done with solids. Crystal formations are beautiful examples of 3-D tessellations. The precise structures of crystals lend themselves to being described and categorized by this mathematics. *Polyhedra, regular polygons, symmetry, tessellations, projections* are a few of the mathematical concepts used to describe properties of crystals. The analysis of crystals began in the 17th century

with work by Johannes Kepler and Robert Hooke. In the early 1900s X-ray crystallography used diffraction diagrams[1] to identify and describe crystals. Since all crystals then known were periodic, the periodicity of crystals was included as part of their definition. Although the periodicity of crystals had not been proven, it became a matter of convention. This meant a crystal had to be composed of a periodic arrangement of identical building blocks (called unit cells). All cells of the crystal had to be formed from the same polyhedron unit. In essence, each and every crystal was considered a 3-D periodic tiling, and periodic tiling then meant the crystal could not possess fivefold symmetry.

It wasn't until the 1980s that a connection was found between Penrose tiles and crystals. They were not connected by something they shared in common, but rather by something they supposedly did not have in

---

1 Patterns created when beams of X-rays or electrons are scattered by the atoms of a solid. A photograph of this illustrates the atomic structure of the solid by the arrangement of white spots on a dark background of the X-ray film. If the spots are fuzzy, not discrete, the solid is glass rather than crystal. Solid matter is either amorphous (without shape) or crystalline. In an amorphous solid, the atoms are randomly arranged with no pattern. A crystal's atoms are arranged with a periodic order to them

common —namely, five-fold symmetry. Until then, it was believed that all crystals were supposedly periodic and thus could not have fivefold symmetry.

In 1982, chemist Daniel Shechtmann discovered a way to join maganese and aluminum to make a new super strong alloy. Upon examining this crystal using X-ray diffraction patterns, he did not find the three, four or six-fold symmetry which crystals were supposed to possess. Instead, he found fivefold[2] symmetry. How could this be? It went against the then known mathematical properties of crystals. However, after two and a half years of trying to uncover errors in his work and then trying to convince fellow scientists that his results were accurate, Shechtmann's findings were made public.

In the meantime, in 1984, physicist Paul Steinhardt and graduate student Don Levine generated a computer simulation composed of virtual 3-D Penrose tiles acting as building blocks of real atoms. They calculated the X-ray diffraction of their imaginary solid. Since the atomic structure of this solid was composed of nonpediodic tiles, they did not expect to see clear bright spots, but rather a blurry diffraction pattern. Instead, the spots were distinct and clear ( like those of crystals), but appeared in a fivefold symmetry pattern. Steinhardt dubbed the solid a *quasicrystal*. It turned out that the computer simulation's diffraction spot pattern resembled that of the aluminum-maganese alloy's— with both exhibiting fivefold symmetry!

These discoveries shook up crystallographers, but it did not invalidate the mathematics of the structure of crystals. Instead, crystallographers

2 Crystallographic restriction or Barlow's law is a mathematical theorem stating that fivefold symmetry is impossible in periodic tiling of a plane or space. Scientists using this theorem in their work apparently had not yet considered non-periodic tiling of crystals possible.

redefined[3] crystals to include these new crystals, the *quasicrystals.* Since Shechtmann's alloy discovery, one quasicrystal has led to another until now there are over a hundred such alloys, some even having 7-fold and 9-fold symmetry.

Mathematics has many such incidents in which the discovery of an idea, proving or disproving a theorem, the solving of a problem does not mean "case closed".[4] Instead, new doors of discovery are open which often shed light on old ideas.

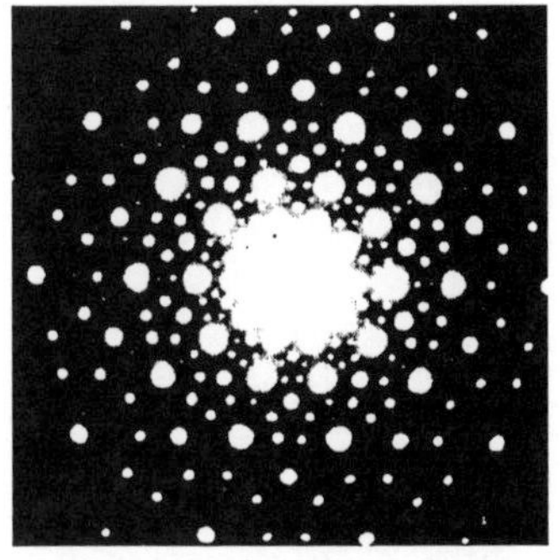

A rendition of the fivefold symmetry as exhibited in the X-ray diffraction patterns of manganese-aluminum alloy quasicrystal.

A rendition of the fivefold symmetry pattern of a sand dollar.

---

3 Until the advent of quasicrystals, only two structural forms of solid were known — crystalline (crystals) and amorphous (e.g. glass, four). Crystals are based on repetitive (i.e. periodic) building blocks, while amorphous solids have no periodic structure. Today, crystals are defined as *any solid with a clear bright discrete diffraction diagram.* The X-ray diffraction diagram for glasses always results in blurred spots. This is why this new definition of crystals also includes quasicrystals.

4 Among these many incidents we have —the attempted proofs of Euclid's Parallel postulate which led to non-Euclidean geometries, Euler's solution to the Königsberg bridge problem which introduced the use of networks in topology, the mathematical monsters of 19th century mathematicians which were the predecessors to fractals, and Fermat's Last Theorem which led to many new mathematical ideas and their interconnections. And the list goes on an on.

# smart machines

Listen, there's a hell of a universe next door;
let's go!

—e.e. cummings

Smart machines look like your everyday handy tools on the surface, but looking inside we find the mathematics of fuzzy logic at work. Such machines are not simply on/off technology, but have built into them special programming and sensors that are designed to analyze the best way to perform a task. In the past, machines were simply made to be turned on when performing a job, such as washing clothes and rinsing for so many minutes. Today, fuzzy logic has changed how such machines work. Consider something as innocuous as a vacuum cleaner. With a dumb vacuum, a manual carpet pile setting is available, in addition to the on/off switch. A smart vacuum is designed to set itself to a carpet's pile. Using infrared sensors, it detects the amount of dirt and dust accumulated and adjusts its suction and brushes as it proceeds across the carpet.

Fuzzy logic's origins date back to 1920 with the work of logician Jan Lakasiewicz, who revised traditional yes-no logic to "multivalent" (multivalued) logic. Yes-no logic uses the numbers 1 and 0 (also referred to as true -false) and Boolean algebra to program and direct tasks. Multivalued logic can consider gray areas by using 0, 1 and every value between them. In 1965, mathematician Lotfi Zadeh applied multivalued logic to set theory, and developed the *fuzzy set,* which was instrumental in the development of the mathematics behind fuzzy thinking. Unlike traditional logic, fuzzy logic, with its multivalued system, has room for "extenuating circumstances" — room for considering many possibilities. Situations for smart machines are not simply assigned *yes-no* values, but rather *if-then.* Data is inputted into the smart machines via its various sensors. This information is then sent to its control system, where. the data is converted into a range of possible values. This process is called *fuzzification.* A smart machine has built into it a set of procedure rules which analyzes these fuzzified values. The system then selects a way to proceed, and creates a set of output values. In order to be used electronically, the output values are *defuzzified* to yes-no control commands.

Smart machines help us take the guesswork out of certain tasks. For example, smart microwave machines have sensors that cook food to a specified temperature, while the dumb microwaves cook solely by the time and temperature you set. Smart washing machines determine the best washing cycle from your input of water level, load size, fabric, dirt, and stain levels. After a wash cycle, the machine will repeat the cycle if it senses the rinse water is not clean enough. Single button smart washing machines have been developed by Matsushita and Hitatchi. You can set smart clothes dryers to sensor dry, rather than selecting a drying time and temperature setting.

*Where are these smart machines?* All around you. — when you enter an elevator, use an autofocus camera, cook something with a sensor microwave oven, drive in your automobile equipped with airbags and other systems, travel by jet plane, or even consult an automated financial advisor.

Here's a partial list of where some smart machines are found:

- Handwriting recognition systems for handheld computers.
- Backlight control in camcorders
- Subway system which senses and adapts ride and power efficiency
- Voice recognition use in development of voice command computers
- Earthquake prediction equipment
- Automobile cruise controls, antilock braking systems, computerized engine efficiency

- Air conditioning systems
- Robotics and robot designs
- Finance assessments
- Automobiles equipped with safety systems which recognize when a driver is falling asleep at the wheel
- Guided missiles

*MATHEMATICAL FOOTPRINTS*

*Previous page:*
*Mathematical problems from a small portion of the Rhind Papyrus of ancient Egypt.*

# Appendix

***How to find that special point on a segment that divides the segment into the golden ratio's value.***

*The golden ratio point divides any segment, call it AC, into the following ratio:* $|AC|/|AB|=|AB|/|BC|$ A B C

Let $|AC| = 1$ unit and $|AB|=x$ unit, then $|BC|=1\text{-}x$. Substituting these values into the ratios above, we get $1/x=x/(1\text{-}x)$ solving for x by using the quadratic formula, the values for x are $(1+\sqrt{5})/\text{-}2$ and$(1\text{-}\sqrt{5})/\text{-}2$. Since x must be positive because it is the length of AB, the positive value for x is used, namely $(1\text{-}\sqrt{5})/\text{-}2$. Replacing this in $|AC|/|AB|$, the golden ratio is $(1+\sqrt{5})/2\approx 1.6180339$ called phi, $\phi$.

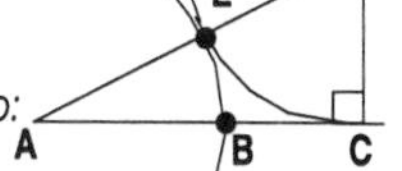


There are various ways to locate phi. *One way is to:*

1) Draw a segment AC
2) Draw CD perpendicular to AC and half its length.
3) Draw AD
4) With center D and radius $|DC|$, draw an arc intersecting AD at E.
5) With center A and radius $|AE|$, draw and arc intersecting AC at B.
6) B's location divides AC into the golden ratio. By letting $|AC|=2x$, then $|CD|$ &$|DE|$ equal x, and $|AD|$ can be obtained using the Pythagorean theorem, namely $|AD|=x\sqrt{5}$,
   Thus, phi$=|AC|/|AB| = 2x/(x(\sqrt{5}\text{-}1)) = 2/(\sqrt{5}\text{-}1) =1.6180339\ldots$

*another way:*

1) Make a square.
2) Find the location of the midpoint, M.
3) Measure the distance from M to corner N, and mark off this distance as shown, MC
4) Complete the rectangle as shown.
5) Point B is the point that divides segment AC into a golden ration.

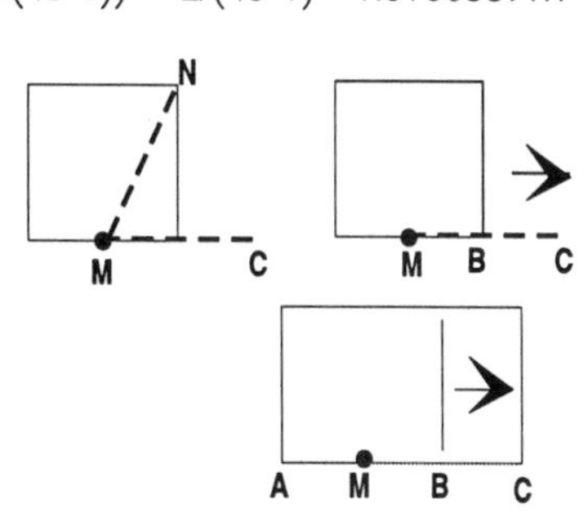

# Appendix

**PREFIXES USED FOR WEIGHTS AND MEASURES**

| **prefix** | **symbol** | **value** |
|---|---|---|
| yotta | Y | 1 000 000 000 000 000 000 000 000 $=10^{24}$ |
| zetta | Z | 1 000 000 000 000 000 000 000$=10^{21}$ |
| exa | E | 1 000 000 000 000 000 000$=10^{18}$ |
| peta | P | 1 000 000 000 000 000$=10^{15}$ |
| tera | T | 1 000 000 000 000$=10^{12}$ |
| giga | G | 1 000 000 000$=10^{9}$ |
| mega | M | 1 000 000$=10^{6}$ |
| kilo | k | $1000=10^{3}$ |
| hecto | h | $100=10^{2}$ |
| deka | da | $10=10^{1}$ |
| | $1=10^{0}$ | |
| deci | d | $0.1=10^{-1}$ |
| centi | c | $0.01=10^{-2}$ |
| milli | m | $0.001=10^{-3}$ |
| micro | µ | 0.000 001$=10^{-6}$ |
| nano | n | 0.000 000 001$=10^{-9}$ |
| pico | p | 0.000 000 000 001$=10^{-12}$ |
| femto | f | 0.000 000 000 000 001$=10^{-15}$ |
| atto | a | 0.000 000 000 000 000 001$=10^{-18}$ |
| zepto | z | 0.000 000 000 000 000 000 001$=10^{-21}$ |
| yocto | y | 0.000 000 000 000 000 000 000 001$=10^{-24}$ |

Ball, W.W. Rouse and Coxeter, H.S.M.. *Mathematical Recreations and Essays,* Dover Publications, Inc., New York, 1987.

Banchoff, Thomas F.. *Beyond The Third Dimension,* W.H. Freeman & Co., New York, 1990.

Barnsley, Micahel. *Fractals Everywhere,* Academic Press, Inc., Boston, 1988.

Bell, Eric Temple. *The Magic Of Numbers,* Dover Publications, Inc., New York, 1946.

Blackwell, William. *Geometry & Architecture,* Key Curriculum Press, Berkeley, CA, 1994.

Boles, Martha and Newman, Rochelle. *Universal Patterns,* Pythagorean Press, Bradford, MA 1987.

Boyer, Carl. *A History of Mathematics,* Princeton University Press, New Jersey, 1985.

Briggs, John and Peat, David. *Turbulent Mirror,* Harper & Row, New York, 1989.

Campbell,D.M. and Higgins, J. editors. *Mathematics: People, Problems, Results,* Wadsworth International, Belmont, CA 1984.

Casto, John L. *Complexification,* Harper Collins, New York, 1994.

Chadwick, John. *Reading the Past—Linear B,* Univ. of California Press, Berkeley, 1987.

Cipra, Barry. *What's Happening in Mathematical Sciences 1995-96,* American Mathematical Society, Providence, RI, 1996.

Cipra, Barry. *What's Happening in Mathematical Sciences 1994,* American Mathematical Society, Providence, RI, 1994.

Cipra, Barry. *What's Happening in Mathematical Sciences 1993,* American Mathematical Society, Providence, RI, 1993.

Cipra, Barry. *What's Happening in Mathematical Sciences 1998-99* American Mathematical Society, Providence, RI, 1999.

Clark, Don. *Myers Laboratories; Hearing in Three Dimensions,* San Francisco Chronicle, San Francisco, June 11, 1987.

Cook, Theodore A.. *The Curves of Life,* Dover Publications, Inc., New York, 1979

Coveney, Peter &Highfield, Roger. *The Arrow of Time,* Fawcett Columbine, New York, 1990.

Dantzig, Tobias. *Number,* MacMillan Co., New York, 1930.

Davies, Paul. *About Time,* Simon & Schuster, New York, 1995.

Davies, Paul. *The Mind of God,* Simon & Schuster, New York, 1992.

Davies, W.V.. *Reading the Past — Egyptian Hieroglyphs,* University of California Press, Berkeley, CA, 1987.

Devlin, Keith. *Mathematics The Science of Patterns,* Scientific American Libarary, New York, 1994.

Dewdney, A.K.. *The Magic Machine,* W.H.Freeman & Co., New York, 1990.

Dewdney, A.K.. *The Armchair Universe,* W.H. Freeman & Co., New York, 1988.

Dilke, O.A.W.. *Reading the Past—Mathematics & Measurment,* University of California Press, Berkeley, CA, 1987.

Dunham, William. *Journey Through Genius,* John Wiley & Sons, New York, 1990.

Eames, charles & Ray. *A Computer Perspective,* Harvard Univ. Press, Cambridge, MA, 1990.

Feynman, Richard P.. *QED,* Princeton Univ. Press, Princeton, NJ, 1985.

Gardner, Martin. *Fractal Music, Hypercards and more,* W.H.Freeman & Co., New York, 1991.

Gardunkel, Solomon & Steen, Lynn A. editors. *For All Practical Purposes,* W.H.Freeman & Co., New York, 1991.

Gleick, James. *Chaos,* Penquin Group, New York, 1987.

Golos, Ellery B.. *Foundations of Euclidean and Non-Euclidean Geometries,* Holt, Rinehart & Winston, Inc., New York, 1968.

Goudsmit, Samuel 7 Clairborne, Robert. *Time,* Time Inc., New York, 1966.

Greensberg, Marvin Jay. *Euclidean and Non-Euclidean Geometries,* W.H. Freeman & Co., New York, 1980.

Grübaum, Branko and Shephard, G.C.. *Tillings and Patterns,* W.H.Freeman and Co., New York, 1987.

Gullberg, Jan. *Mathematics from the Birth of Numbers,* W.W. NORTON & CO., New York, 1997.

Hall, Stephen S. *A Molecular Code Links Emotions,* Smithsonian Magazine, Smithsonian Asso., Washington D.C., 1990.

Hambridge, Jay. *The Elements of Dynamic Symmetry,* Dover Publications, Inc., New York, 1967.

Hawkins, Gerald s.. *Mindsteps to the Cosmos,* Harper & Row, Publishers, NY, 1983.

Hoffman, Paul. *Archimedes' Revenge,* W.W.Noron & Co., Inc., New York, 1988.

Ifrah, Georges. *From One to Zero,* Penquin Books, New York, 1985.

Ivins, Jr., William M.. *Art & Geometry,* Dover Publications, Inc., New York, 1946.

Kaku, Michio. *Hyperspace,* Oxford University Press, New York, 1994.

Kosko, Bart. *Fuzzy Thinking,* Hyperion, New York, 1993.

Kosko, Bart. *Fuzzy Future,* Harmony Books, New York, 1999.

Krauss, Lawrence M.. *The Physics of Star Trek,* Basic Books/Harper Collins, New York, 1995.

Lebow, Irwin. *The Digital Connection,* Computer Science Press, New York, 1991.

Lewin, Roger. *Complexity,* Macmillan Publishing Co., New York, 1992.

Lloyd, Steon; Rice, David; et al. *World Architecture,* McGraw-Hill Co., London, 1978.

Luckiesh, M.. *Visual Illusions,* Dover Publications, Inc., New York, 1965.

Macrone, Michael. *Eureka!,* Harper Collins , New York, 1994.

Macvey, John W. *Time Travel,* Scarborough House/Publishers, Chelsea, MI, 1990.

Mandelbrot, Benoit. *The Fractal Geometry of Nature,* W.H.Freeman & Co., New York, 1983.

Menninger, Karl. *Number Words 7 Number Symbols,* Dover Publications, New York, 1969.

Newman, James R., et. al.. *The World of Mathematics,* Simon & Schuster, New York, 1956.

Nicolis, Grégoire and Prigogine, Ilya. *Exploring Complexity,* W.H. Freeman and Co., New York, 1989.

Pagels, Heinz R.. *Perfect Symmetry,* Simon & Schuster, New York, 1985.

Palfrreman, Jon and Swade, Doron. *The Dream Machine,* BBC Books, London, 1991.

Pappas, Theoni. *More Joy of Mathematics,* Wide World Publsihing/Tetra, San Carlos, CA, 1991.

Pappas, Theoni. *The Joy of Mathematics,* Wide World Publishing/Tetra, San Carlos, CA, 1989.

Pappas, Theoni. *What Do You See?—An optical illusion slide show,* Wide World Publishing/Tetra, San Carlos, CA, 1989.

Paulos, John Allen. *Beyond Numeracy,* Alfred A. Knopf, New York, 1991.

Peacock, Roy E.. *A Brief History of Eternity,* Crossway Books, Wheaton, IL, 1990.

Peat, E. David. *Superstrings & the Search for TOE,* Contemporary Books, Chicago, 1988.

Pedoe, Dan. *Geometry and the Visual Arts,* Dover Publications, Inc., New York, 1976.

Peitgen, H., Jürgens, H., & Saupe, Dietmar. *Chaos & Fractals,* Speinger-Verlag, New York, 1992.

Peterson, Ivars. *The Jungle of Randomness,* John Wiley & Sones, Inc., New York, 1990.

Peterson, Ivars. *Fatal Defect,* Random House, New York, 1995.

Pickover, Clifford. *Computers and the Imagination,* St. Martin's Press, New York, 1991.

Pickover, Clifford. *Mazes of the Mind,* St. Martin's Press, New York, 1992.

Pierce, John R.. *The Science of Musical Sound,* W.H.Freeman & Co., New York, 1983.

Prigogine, Ilya. *Order Out Of Chaos,* Bantam Books, Toronto, 1988.

Ransom, William R.. *3 Famous Geometries,* William Ransom, 1959.

Regis, Ed. *Nano,* Little, Brown & Co., Boston, 1995.

Rheingold, Howard. *Virtual Reality,* Summit Books, New York, 1991.

Rhoribuchi, Seiji, editor. *Stereogram,* Masahiro Oga, Japan, 1994.

Richter, Jean Paul. *The Notebooks of Leonardo da Vinci,* Dover Publications, New York, 1970.

Robbin, Tony. *FOURFIELD:Computers, Art and the 4th Dimension,* Little Brown & Co., Boston, 1992.

Robins, Gay and Shute, Charles. *The Rhind Mathematical Papyrus,* Dover Publications, Inc., New York, 1987.

Ronan, Colin A.. *Science,* Facts on File Publications, New York, 1983.

Rucker, Rudy. *The Fourth Dimension,* Houghton Mifflin Co., Boston, 1984.

Ruelle, David. *Chance and Chaos,* Princeton Uninersity Press, New Jersey, 1991.

Schroeder, Manfred. *Fractals, Chaos, Power Laws,* W.H.Freeman and Co., New York, 1990.

Shlain, Leonard. *Art & Physics,* William Morrow and Co., New York, 1991.

Singh, Simon. *Fermat's Enigma,* Walter & Co., New York, 1997.

Smith, D.E.. *History of Mathematics,* Dover Publications, Inc., New York, 1951.

Smoot, George & Davidson, Keay. *Wrinkles in Time,* William Morrow & Co. inc., New York, 1993.

Sobel, Dava. *Longitude,* Pengiun Books, New York, 1995.

Stevens, Peter S.. *Patterns In Nature,* Little Brown & Co., Boston, 1974.

Stewart, Ian and Golubitsky, Martin. *Fearful Symmetry-Is God a Geometer?* Blackwell Publishers, Oxford, 1992.

Stewart, Ian. *Does God Play Dice?* Basil Blackwell, Oxford, 1989.

Struik, Dirk. *A Concise History of Mathematics,* Dover Publications Inc., New York, 1967.

Stwertka, Albert. *Recent Revolutions in Mathematics,* Franklin Watts, New York, 1987.

Waerden, B.L. van der. *Science Awakening,* John Wiley & Sons, Inc., New York, 1963.

Waldrop, M. Mitchell. *Complexity,* Simon & Schuster, New York, 1992.

Walker, C.B.F. *Reading the Past —Cuneiform,* Univ. of California Press, Berkeley, 1987.

Wells, David. *Curious & Interesting Numbers,* Penguin Books, London, 1986.

Wolf, Fred Alan. *Parallel Universes,* Simon & Schuster, New York, 1988.

Articles from the following periodicals:
*Science News, Look Japan, Scientific American, Discover, Science,* and *Notices (American Mathematical Society.*

## About the Author

Mathematics teacher and consultant Theoni Pappas received her B.A. from the University of California at Berkeley in 1966 and her M.A. from Stanford University in 1967. Pappas is committed to demystifying mathematics and to helping eliminate the elitism and fear often associated with it.

Her innovative creations include *The Mathematics Calendar, The Math-T-Shirt, The Children's Mathematics Calendar, The Mathematics Engagement Calendar,* and *What Do You See?*—an optical illusion slide show with text. Pappas is also the author of the following books: *Mathematics Appreciation, The Joy of Mathematics, Greek Cooking for Everyone, Math Talk, More Joy of Mathematics, Fractals, Googols and Other Mathematical Tales, The Magic of Mathematics, The Music of Reason, Mathematical Scandals, The Adventures of Penrose —The Mathematical Cat* and *Math for Kids & Other People Too!*. Her latest books, published in 1999, are *Math-A-Day* and *Mathematical Footprints* .